ANNALES ACADEMIÆ SCIENTIARUM FENNICÆ

MATHEMATICA

DISSERTATIONES

141

EQUIVARIANT TRIANGULATIONS OF DIFFERENTIABLE AND REAL-ANALYTIC MANIFOLDS WITH A PROPERLY DISCONTINUOUS ACTION

TUOMAS KORPPI

University of Helsinki, Department of Mathematics and Statistics

To be presented, with the permission of the Faculty of Science of the University of Helsinki, for public criticism in Auditorium CK112, Exactum, on May 20th, 2005, at 12 o'clock noon.

HELSINKI 2005
SUOMALAINEN TIEDEAKATEMIA

ISSN 1239-6303
ISBN 951-41-0973-2

Received 14 April 2005

2000 Mathematics Subject Classification:
Primary 57R05; Secondary 57S30, 32C05.

YLIOPISTOPAINO
HELSINKI 2005

I thank Sören Illman for contributing the conjecture that a real-analytic manifold with a properly discontinuous real-analytic action admits an equivariant triangulation that is at the same time subanalytic and C^r for any chosen $r < \infty$ (where the triangulation may depend on r. Note that my final result is somewhat stronger than the original conjecture.)

I thank my fellow graduate students Roman Goebel, Anssi Lahtinen, Pekka Pankka, Juha-Matti Perkkiö and Elena Ravaioli for helpful discussions.

I thank the Magnus Ehrnrooth Foundation, the Graduate School of Analysis and its Applications and Helsingin yliopiston tiedesäätiö (The Research Foundation of the University of Helsinki) for funding my research.

I thank Tuula Blåfield for checking the language and Tuulikki Mäkeläinen for helping with the arrangements needed to print this thesis.

Contents

0.1 Introduction

In this thesis we prove the following theorem

Theorem 1 *Let (M,N) be a C^r-manifold pair (M,N), $1 \le r \le \omega$, with a properly discontinuous C^r-action of G (an arbitrary discrete group).*

1. *Then (M,N) has a G-equivariant C^r-triangulation. Here $1 \le r < \infty$ denotes the differentiable case, $r = \infty$ denotes the smooth case, and $r = \omega$ denotes the real analytic case.*

2. *The triangulation is in a certain sense unique: Assume that we have two G-equivariant C^r-triangulations $t_1 \colon |K_1, L_1| \to (M,N)$, and $t_2 \colon |K_2, L_2| \to (M,N)$. Then there exist equivariant subdivisions (K_1', L_1') and (K_2', L_2') of (K_1, L_1) and (K_2, L_2), respectively, and G-equivariant C^1-approximations $t_1' \colon |K_1', L_1'| \to (M,N)$ and $t_2' \colon |K_2', L_2'| \to (M,N)$ of t_1 and t_2, respectively, such that $(t_2')^{-1} \circ t_1' \colon (K_1', L_1') \to (K_2', L_2')$ is a G-equivariant simplicial isomorphism.*

   Furthermore, t_1' is G-homotopic to t_1 and t_2' is G-homotopic to t_2. For $1 \le r \le \infty$ the homotopy is of the class C^r, and for $r = \omega$, the homotopy is of the class C^∞.

By a C^r-triangulation we mean a triangulation where the embeddings of closed simplices into the manifold are C^r-immersions.

We actually prove the result for a slightly wider class of actions than properly discontinuous actions. See Theorem 128 for the exact formulation of the result. We also prove that proper discontinuity is an almost necessary condition for the existence of triangulation. That is, it is a necessary condition in the class of connected manifolds with an effective action. This result allows us to find an *if and only if* condition for the existence of a triangulation in the class of C^r-manifold pairs with a C^r-action of a discrete group. (See Theorem 128.)

We prove these results both for manifolds without a boundary and for manifolds with a boundary. In the latter case we make some requirements for the compatibility of ∂M and N.

This thesis is an expanded and improved version of my licentiate thesis [15], where I proved the same results for the class of C^r, $1 \leq r \leq \infty$, manifolds with a properly discontinuous action, but I omitted the real analytic case and the generalization from the class of properly discontinuous actions.

Our results extend the triangulation results in Munkres [18] for non-equivariant C^r-manifolds (originally proved by Whitehead [24]) and also the results by Illman [7] for C^∞-manifolds with an action of a finite group.

The basic idea of our proof is taken quite directly from Munkres [18] and Illman [7]. However, contrary to the situation of Munkres [18] and Illman [7], in our case it is not always possible to embed the manifold pair into a linear space with a linear action. This forces us to use local coordinates in our manifolds, and as a consequence we face some technical problems not present in the setting of Illman [7] and Munkres [18].

I do not know how to construct an equivariant collar or an equivariant double for a real analytic manifold with a boundary. Thus, I cannot use Illman's triangulation technique for manifolds with a boundary, where one first triangulates the double and then restricts the triangulation into one half only. I cannot use Munkres' technique either, where one triangulates the collar and the interior of the manifold separately and then fits the triangulations together. I triangulate a manifold with a boundary in a direct way, which causes some changes to the traditional ways.

The major difference between the C^∞-case and and C^r-case, $r < \infty$ is that the tangent bundle map induced by a C^∞-map is still C^∞, but for a C^r-map it is only C^{r-1}. Because of this, I have chosen a quite elementary approach, and I do not use the elaborate device of a tangent bundle. This approach is used when constructing coordinate neighbourhoods.

Many preliminary results in this thesis are rather trivial modifications of well-known results. However, in the literature one can find proofs only for the C^∞-case (and not for C^r-case, $r < \infty$) or proofs that do not work as such in the setting where we have a properly discontinuous action. As a consequence, I give proofs of elementary facts in our specific situation. Sometimes I did not need to make many changes to existing proofs, in other cases I had to dig into sources dating from the 1940's and make appropriate changes in them. In some other cases I had to invent proofs myself almost from a scratch.

In Chapter 1, we study an equivariant substitute for coordinate neighbourhoods. In Chapter 2, we study equivariant C^r-maps from simplicial complexes into manifolds and C^1-approximations of such maps. In Chapter 3, we study the particular approximations that are needed in constructing an equivariant triangulation and in the uniqueness result.

In Chapter 4 we prove the main results of this thesis. In sections 4.2 and 4.4 we use the techniques of Munkres and Illman to prove the existence and uniqueness of triangulation, but due to the use of local coordinates, we had to make quite drastic modifications. For Section 4.5, where I prove the almost necessity of proper discontinuity, we had no preexisting model to follow.

We need to use the twisted product construction also for abstract simplicial complexes. The notion of a simplicial twisted product is a straight-forward modification of the notion of a topological twisted product. However, as far as I know, it has not been developed anywhere in the literature, and thus we show some essential results in Appendix A.

We assume that the reader is familiar with the basics of differentiable manifolds, topological transformation groups and simplicial complexes.

The most of this thesis is simply following footsteps of Munkres [18] and Illman [7], and the major contribution of this thesis is to write down a generalization of their results explicitly. However, there are some minor things that I am in particular proud of. They are the following:

- Finding out a way to triangulate manifolds with a boundary without constructing a collar.
- Finding the old results of Bochner [2] (Lemma 28 in this thesis) and Whitehead [24] (Lemma 87 in this thesis), which help us to obtain a triangulation of the right C^r-class.
- Introducing the concept of a twisted product of abstract complexes, which enables us to prove results about equivariant subdivisions without resorting to hand-waving. (Appendix A)
- The result that proper discontinuity is an almost necessary condition for the existence of triangulation. (Section 4.5)
- Useful lemmas 36, 11 and 13.

0.2 History

The result that a C^r-manifold, $1 \leq r \leq \omega$, admits a C^r-triangulation unique up to subdivision and arbitrarily small changes is due to Whitehead [24]. Actually, the main body of Whitehead's paper deals only with C^1-triangulations, but at the end of the paper he mentions that his construction generalizes also for the C^r-case, $1 \leq r \leq \omega$.

Since Whitehead's notation is slightly outdated and thus difficult to read, it is the author's impression that the standard reference to Whitehead's result is Munkres [18].

However, Munkres made one crucial change to Whitehead's argument so that his version works only for $1 \leq r \leq \infty$. (See the Extension Lemma 87 of this thesis and compare it to Munkres [18], Lemma 9.8.) A modern exposition of the real-analytic case appeared in Shiota [22]. Shiota made the Extension Lemma in a similar way to Whitehead, but for the rest of the construction, he did not do things from the scratch, but rather reduced the existence of the C^ω-triangulation to the C^∞-case.

Apart from triangulations of differentiable manifolds, work has been done on triangulations of semi-algebraic, semi-analytic and subanalytic sets. We know that such sets admit a triangulation in the respective category unique up to subdivision. The semi-analytic case is due to Giesecke [5] and Lojasiewicz [16], and the semi-algebraic and subanalytic cases are due to Hironaka [4], building on Lojasiewicz' ideas.

The subanalytic triangulation result is rather strong, since the embeddings of *open* simplices into the triangulated space are real-analytic embeddings. However, in the general case one cannot hope to get a real-analytic triangulation in the sense of this thesis of a subanalytic set, since that would mean that the embeddings of *closed* simplices to the space should be immersions, and, consequently, there should not be corners of the zero angle. General subanalytic (or even semi-algebraic) sets do not admit such a triangulation. Consider, for example, the set

$$\{(x, y) \in \mathbb{R}^2 \mid x, y \geq 0, y \leq x^2, x \leq 1\}.$$

The techniques of Lojasiewics and Hironaka can be extended to obtain a subanalytic triangulation of a real-analytic manifold, which is unique up to subdivision. We do not know if one can avoid corners of zero angle of simplices in manifolds by Lojasiewics' and Hironaka's techniques.

However, there is no hope to get a real-analytic triangulation of a manifold (in the sense of this thesis), which is unique up to subdivision (versus unique up to subdivision and arbitrarily small changes.) Namely, it is possible to construct two different real-analytic triangulations of $\mathbb{R}^2$, such that all their common subdivisions have corners of zero angle. Let $t\colon |K| \to \mathbb{R}^2$ be a triangulation of $\mathbb{R}^2$ such that $t^{-1}\{(x, 0) \mid 0 \leq x \leq 1\}$ is a 1-simplex of $|K|$. Furthermore, let $f\colon (x, y) \mapsto (x, y + x^2)$ be a real-analytic automorphism of $\mathbb{R}^2$. Then $f \circ t$ is a triangulation of $\mathbb{R}^2$ such that all common subdivisions of t and $f \circ t$ have corners of zero angle.

There seem to be two different ways to extend the triangulation results to the equivariant case.

The first is to follow Munkres' construction from the beginning to the end and at each step make things equivariantly. This path was followed by Illman [7], where he proved that each C^∞-manifold with a C^∞-action of a finite group has a unique (up to subdivision and arbitrarily small changes) equivariant triangulation. With some changes (the major change being that we replace the Munkres' simplified Extension Lemma by the Whitehead's original), Illman's techniques work also for all C^r, $1 \leq r \leq \omega$, as we prove in this thesis.

The second way works for real-analytic manifolds with a proper real-analytic action of a Lie group. It is known that we can equip the orbit space with a subanalytic structure and then triangulate the orbit space using subanalytic triangulation results. This is due to Matumoto and Shiota [17] (see also Kankaanrinta [11]).

If the acting Lie group is zero-dimensional (for zero-dimensional Lie groups, proper actions coincide with properly discontinuous actions) one can lift the simplices of the orbit space into simplices of the original space, and obtain an equivariant and unique subanalytic triangulation of the original space. If the acting Lie group has one dimension or more, one obtains a triangulation of the original space in a generalized sense (see Illman [9]). We do not know whether one can use these techniques so that the lifted (open or closed) simplices are real analytic. At least it is not obvious by the works of Matumoto, Shiota and Illman.

0.3 Preliminaries and conventions

In this section we state some well-known basic results and fix the notation.

Differentiability

A map $f\colon \mathbb{R}^m \to \mathbb{R}^n$ is C^r, $1 \leq r < \infty$, if f has all continuous partial derivatives up to the degree r. The map f is C^∞ or smooth if it is C^r for all $1 \leq r < \infty$. The map f is real-analytic, or C^ω, if it coincides with a Taylor series in a neighbourhood of any point.

Since f being C^ω implies that f is C^∞, we make a notational exploitation and assume that $\omega > \infty$. This allows us to write expressions like f is C^r, $1 \leq r \leq \omega$.

A manifold with an atlas (that is, a collection of local coordinate systems such that each point of the manifold is contained in some coordinate system) ψ is of a class C^r if the change-of-coordinates functions between charts are of the class C^r. We usually suppress the atlas from notation. A map between manifolds is C^r, if it is C^r in local coordinates.

Let G be a discrete group. We say that (M, N) is a G-equivariant C^r-manifold pair if G acts on M by C^r-diffeomorphisms, and N is a closed invariant submanifold of M such that $\partial N \subset \partial M$, and the embedding of N into M is of the class C^r. We futher assume that $T_xN \not\subset T_x\partial M$ for all $x \in N \cap \partial M$. Since the codimension of ∂M is 1, the last assumption is equivalent to the assumption that ∂M is transversal (defined below) to T_xN everywhere in $N \cap \partial M$.

Throughout this thesis, (M, N) denotes a C^r-manifold pair, $1 \le r \le \omega$, possibly with a boundary, and G is an arbitrary fixed discrete group acting properly discontinuously and in a C^r-way on (M, N).

We will work with a fixed r and mean by diffeomorphic C^r-diffeomorphic.

We denote $\mathbb{R}_+ = [0, \infty[$, and $H^n = \{(x_1, \dots, x_n) \in \mathbb{R}^n | x_1 \in \mathbb{R}_+\}$.

If $x \in M$ and $f\colon N \to M$, $f'\colon N' \to M$ are embeddings such that $x \in f(N) \cap f'(N')$, we say that f and f' are transversal at x if $D_{f^{-1}(x)}f \oplus D_{f'^{-1}(x)}f'$ is a surjective map $T_{f^{-1}(x)}N \oplus T_{f'^{-1}(x)}N' \to T_xM$.

There is a well-known result called the inverse function theorem, which states the following: Let $U \subset \mathbb{R}^n$ be open and $f\colon U \to \mathbb{R}^n$ a C^r-function such that for $x \in U$ we have that D_xf is a bijection $T_xU \to T_{f(x)}\mathbb{R}^n$. Then there exists a neighbourhood U' of x such that $f|U'$ is a diffeomorphism. For the case $1 \le r \le \infty$, this is Corollary 1.12 in Munkres [18]. The $r = \omega$-case is proved for example in Narasimhan [19]. Narasimhan states the implicit function theorem for complex analytic functions (without proof) and reduces the real-analytic inverse function theorem to the implicit function theorem for complex analytic functions. (See 1.3.5 through 1.3.10 in Narasimhan [19].) For the proof of the implicit function theorem for complex analytic functions, see, for example, Kaup-Kaup [12], Theorem 8.6. In the sequel we will use the inverse function theorem without further notification.

To avoid pathological cases, we assume that all our manifolds are paracompact. However, there is no need to assume them to be N_2 (meaning second countable), since the proofs of our results for paracompact manifolds are essentially as easy as in the special case of N_2-manifolds. This is a consequence of the well-known fact that a manifold is paracompact if and only if each of its connected components is N_2. (This fact is proved as follows: A manifold is regular Hausdorff, and a regular Hausdorff N_2-space is metrizable by the Urysohn Metrization Theorem and thus paracompact.

A connected component of a manifold is a manifold. To prove the reverse, it is thus enough to prove that a connected paracompact manifold M is N_2. Let (U_i) be a locally finite open cover of M with relatively compact open sets homeomorphic to an open subset of $\mathbb{R}^n$. Since (U_i) is a locally finite cover with relatively compact sets, it is star-finite. Let $U^1 = \{U_1\}$, and inductively let U^n be the set of those U_i that intersect some $U_j \in U^{n-1}$. Since M is connected, $(U_i) = \bigcup_n U^n$. But each U^n is finite, and thus (U_i) is countable. Since each U_i is a coordinate neighbourhood, it is N_2. Thus M is a union of a countable number of open N_2-subspaces, hence it is N_2.)

Transformation groups

We denote by $G \curvearrowright X$ that G is a topological group acting on a space X. (Since we deal only with properly discontinuous actions, the role of the topology of G is rather limited, cf. the following definition.)

Following Abels–Strantzalos [1], Definition 2.1.2., we define proper discontinuity.

Definition 2 *An action $G \curvearrowright X$ is properly discontinuous if G is a discrete group, and given any two points $x, y \in X$, there exist neighbourhoods U_x and U_y of x and y, respectively, such that $U_x \cap gU_y \neq \emptyset$ for only finitely many $g \in G$.*

It is an immediate consequence of the definition that if $G \curvearrowright X$ properly discontinuously, then G_x is a finite subgroup of G for every $x \in X$. For those readers who are familiar with proper actions, we remark that properly discontinuous is a special case of proper, when the acting group has the discrete topology.

If $G \curvearrowright X$ and $U \subset X$, we denote $G_U = \{g \in G \mid gU = U\}$.

We say that an action $G \curvearrowright X$, where $X \subset \mathbb{R}^n$, is linear (orthogonal) if each $g \in G, g \colon X \to X$ is a restriction of a linear (orthogonal) map. (We often identify the element g of the acting group with the map $X \to X$ that corresponds to it.)

If $G \curvearrowright X$, we denote the orbit space of X by X/G and the natural projection $X \to X/G$ by π.

If X has a C^r-structure, we assume that $g \colon X \to X$ is C^r for all $g \in G$.

In this thesis, if not stated otherwise, each action is properly discontinuous. Furthermore, "a map" means a C^r-differentiable function. (Later on in this thesis, we define when a map from a geometrical realization of a simplicial complex into a manifold is C^r-differentiable.)

Topology

By a slightly unstandard convention, we declare that $A \subset\subset X$ means $\bar{A} \subset \operatorname{int} X$. By a typical Finnish convention, we use the expression N_2 to denote *second countable.*

Since systematical notation is easier to read than arbitrary symbols, we often use $f|$ to mean a restriction of f to some subset. When the restriction subset is evident from the context, we may leave it unspecified in the notation $f|$. In complicated constructions we often define a map f first in some subset A and extend it later to a larger domain. In these cases, we use the notation $f|A$ before we have defined the entire f.

If A is a topological space and B is a subset of A, we say that B is componentwise compact if $B \cap U$ is compact for each component U of A. If B is a collection of objects (simplices, etc.) that lie in A, we say that B is componentwise finite if each component of A contains only finitely many objects belonging to B.

Every (paracompact) manifold is known to be metrizable, and by Abels-Strantzalos [1], Theorem 5.3.2, we can choose an invariant metric for a locally compact metrizable space with a properly discontinous action and a paracompact orbit space. By Abels-Strantzalos [1], Theorem 3.2.5.d, the orbit space for a paracompact manifold is paracompact. Thus, in our situation, when we have a properly discontinous action in a paracompact manifold, we find an invariant metric for the manifold. Actually, we could choose a Riemannian metric as well, but it turns out that this does not help us.

We will often refer to a distance in a manifold. In such sitations we always assume that we have chosen some fixed invariant metric for the manifold.

Lemma 3 *Assume $G \curvearrowright X$ properly discontinuously, and X has a G-invariant metric d. Then, for each orbit O there exists a positive constant c such that for all $y, x \in O$ we have $d(x, y) > c$.*

Proof: Assume on the contrary, that there is a sequence of pairs of points $(x_i, y_i)_{i \in \mathbb{N}}$, all in one orbit, such that $d(x_i, y_i) \to 0$. By replacing each (x_i, y_i) with a suitable $(g_i x_i, g_i y_i)$, we can assume that $x_1 = x_2 = \ldots$. The sequence y_i converges to x_1, that is, any neighbourhood of x_1 contains infinitely many of the points y_i. Now we observe that each y_i equals $g'_i y_1$ for some $g'_i \in G$.

Thus, every neighbourhood of x_1 contains $g'_i y_1$ for infinitely many g'_i, which is impossible for a properly discontinuous action. □

Lemma 4 *If $G \curvearrowright X$ properly discontinuously, and U is a neighbourhood of x, then there exists a G_x-invariant neighbourhood $V \subset U$ of x.*

Proof: Since G_x is finite, we can choose $V = \bigcap_{g \in G_x} gU$.□

A homeomorphism means a surjective homeomorphism, and we use the expression *topological imbedding* to denote a continuous injective map such that the topology of the domain coincides with the relative topology of the image.

To simplify notation, a family of the objects U_i is denoted by (U_i), and the index set is usually not mentioned. However, we assume that the index set of indices i is I and the index set of indices j is J.

We say that a covering (U_i) of X is a G-covering if each of its members U_i is G-invariant.

Simplicial complexes

Following well-known standard definitions, we make a difference between abstract simplicial complexes and their geometrical realizations. An abstract simplicial complex $(K, \mathcal{S})$ is a pair where K is a set of vertices and $\mathcal{S} \subset \mathcal{P}(K)$ is a set of simplices satisfying the following conditions

- $\mathcal{S}$ contains all unit sets.

- Each $s \in \mathcal{S}$ is a non-empty finite set.

- If $s \in \mathcal{S}$ and $s' \subset s$ is non-empty, then $s' \in \mathcal{S}$.

We often suppress $\mathcal{S}$ from the notation and denote the simplicial complex by K.

A geometrical realization $|K|$ of K is a subset of $\Pi_K[0,1]$ consisting of those $(x_v)_{v\in K}$ for which $\sum x_v = 1$ and $\{v \mid x_i \neq 0\}$ is a simplex of K. If s is a simplex of K, then the geometrical simplex $|s|$ is the set of those (x_v) for which $\{v \mid x_i \neq 0\} \subset s$. When topologizing $|K|$, we topologize each geometrical simplex $|s|$ such that it is naturally homeomorphic with the standard simplex (of an adequate dimension), and $|K|$ has the weak topology induced by the topologies of geometrical simplices. We are in particular interested in locally finite simplicial complexes K (i.e. of such K that each $v \in K$ belongs only to a finitely many simplices). In this case, the weak topology (the one defined above) conicides with the topology given by a natural metric (defined below.)

Occasionally, we may identify the geometrical realization $|K|$ of a simplicial complex K with the complex K itself. If X is a topological space, a map $t\colon K \to X$ means the map $t\colon |K| \to X$.

If K is a simplicial complex, we say that a subcomplex L of K is a full subcomplex of K if a simplex s of K is a simplex of L whenever all vertices of s are in L.

If s is a simplex of K, we define the open simplex $|\overset{\circ}{s}| = |s| \setminus \{|b| \mid b$ is a proper face of $s\}$. The simplicial boundary $\overset{\bullet}{s}$ is the subcomplex of s consisting of all proper faces of s.

If $t\colon |K| \to X$ and $t'\colon |K'| \to X$ are topological embeddings of simplicial complexes into a topological space, we say that K and K' intersect in a subcomplex, if the following conditions hold:

- For all $n \in \mathbb{N}$ and all n-simplexes s of K such that $t|\overset{\circ}{s}|$ and $t'|K'|$ intersect, we have that $t'^{-1}t||s|$ is a linear isomorphism $|s| \to |s'|$, where s' is an n-simplex of K'.

- For all $n \in \mathbb{N}$ and all n-simplexes $|s'|$ of $|K'|$ such that $t'|\overset{\circ}{s'}|$ and $t|K|$ intersect, we have that $t^{-1}t'||s'|$ is a linear isomorphism $|s'| \to |s|$, where s is an n-simplex of K.

We say that K and K' intersect in a full subcomplex if, furthermore, $t^{-1}t'|K'|$ is a full subcomplex of K and $t'^{-1}t|K|$ is a full subcomplex of K'.

We remark that if L is a subcomplex of K, then the barycentric subdivision of L is a full subcomplex of the barycentric subdivision of K.

If L is a subcomplex of K, then $\operatorname{St} L$ is the star of L, that is the subcomplex of K that consists of all simplexes s (together with their faces) of K that intersect L. The (simplicial) interior of $\operatorname{St} L$ is denoted by $\overset{\circ}{\operatorname{St}}\, L$, that is $|\overset{\circ}{s}| \subset \overset{\circ}{\operatorname{St}}\, L$ if and only if $|s| \cap L \neq \emptyset$. The n times iterated star operation is denoted by $\operatorname{St}^n L$. If $A \subset |K|$, we denote by $\operatorname{St} A$ the subcomplex that consists of all closed simplexes s (together with their faces) such that $|s| \cap A \neq \emptyset$. The subcomplex $\overset{\circ}{\operatorname{St}}\, A$ is defined to be the set of open simplices $|\overset{\circ}{s}|$ such that $|s| \cap A \neq \emptyset$.

An action $G \curvearrowright K$ means an action on the set of vertices of K such that for each $g \in G$ and each simplex s of K, we have that gs is a simplex of K. Using such an action we obtain an action $G \curvearrowright |K|$ as we extend the action linearly to each geometrical simplex.

If we are given an action of G on a simplicial complex K, we define the orbit complex K/G. Its vertices are the orbits of vertices of K under the action of G. Let $\Pi\colon K \to K/G$ be the orbit map. The simplices of K/G are the sets of the type Πs, where s runs through the simplices of K.

It may occasionally happen that $|K/G|$ and $|K|/G$ do not coincide, and also that the fixed point set of $|K|$ may not consist of entire simplices. However, if K'' is a second barycentric subdivision of K, then we always have that $|K''/G|$ and $|K''|/G$ do coincide, and the fixed point set of $|K''|$ consists of entire simplices. (See Bredon [3], Section 3.1 for proofs.) Since in triangulations a second barycentric subdivision of K is as good as K itself, we may assume that when we are dealing with orbit spaces and fixed point sets, we always switch to the second barycentric subdivision.

An equivariant simplex of a simplicial complex K with an action of G means a subcomplex of K of the type Gz, where z is a simplex of K. We usually denote both usual simplices and equivariant simplices with lower case letters s, s', z, z' and so on.

If s is an equivariant simplex, we say that s is a regular equivariant simplex if the following holds: If $g \in G$ and s' is a simplex of s, then either $gs' \neq s'$ or g fixes s' pointwise. It is enough to consider only regular equivariant simplices, since in the barycentric subdivision of a simplicial G-complex all equivariant simplices are regular.

Let K be a locally finite simplicial complex. There is a natural way to give a simplexwise linear metric for $|K|$, namely, represent each point $x \in |K|$ as a linear combination $\Sigma_v a_v v$, where v runs over the vertices of K and the coefficients are non-negative reals such that $\Sigma_v a_v = 1$. Then $d(\Sigma_v a_v v, \Sigma_v a'_v v) = (\Sigma_v (a_v - a'_v)^2)^{1/2}$. If there is a simplicial action of G in $|K|$, this metric is invariant, and it is linear in the sense that each simplex is isometric with a rectilinear region of some $\mathbb{R}^n$ equipped with the Euclidean metric. Since we assume K to be locally finite, the metric topology of $|K|$ coincides with the weak topology.

We will make some slightly non-obvious subdivisions. Thus we define what we exactly mean by a subdivision. Let K be a simplicial G-complex. An equivariant subdivision of K is a simplicial G-complex K' together with a G-equivariant homeomorphism $t\colon |K'| \to |K|$ such that

- For each simplex s' of K' there exists a simplex s of K such that $t|s'| \subset |s|$.
- For each simplex s' of K' we have that $t||s'|$ is an affine map.

We often suppress the map t from the notation and simply say that K' is a subdivision or a G-equivariant subdivision of K. We identify $|K'|$ and $|K|$ with the homeomorphism

t. When we talk about a distance in $|K'|$, we assume that $|K|$ has a metric, which $|K'|$ inherits by the identification.

Lemma 5 *Let K be a finite simplicial complex with a metric. Then in the iterated barycentric subdivision, the maximum diameter of the simplices converges to 0. Furthermore, if $t\colon K \to M$ is continuous and M is equipped with a metric, then the maximum diameter of t-images of simplices converges to 0.*

Proof: In Spanier, [23], Corollary 3.3.13, the first part is proved for complexes with a metric that is linear in each simplex.

We prove the last part for $|K|$ with a linear metric. Let $\epsilon > 0$ be arbitrary. Let (U_i) be a covering of M with ϵ-balls. Now $(t^{-1}U_i)$ has a finite subcovering, which has a Lebesgue number ϵ'. Now, by the above reference to Spanier, we have an iterated subdivision K' such that each simplex of K' has a diameter less than ϵ'. Now K' is a subdivision such that the diameter of ts is less than ϵ for all $s \in K'$. Thus the last part is proved.

The first part for an aribitrary metric follows, as one uses the result above to the natural simplicial isomorphism $t'\colon |L| \to |K|$, where L is the same complex as K but equipped with a linear metric. The last part for an arbitrary metric follows, since it does not depend on the particular choice of the metric of $|K|$.□

Linearity

For most of the time, we use the term "linear" in the affine sense.

If M is a manifold, $U' \subset \mathbb{R}^n$ and $h\colon U' \to U \subset M$ is a coordinate map, we say that U has the linear (affine) structure defined by $U' \subset \mathbb{R}^n$.

Acknowledgements

Except perhaps for Lemma 3 and some of the more extraordinary conventions, all the results stated in this section are old and well-known basic results in the respective fields of study.

Chapter 1

Local coordinates

Manifolds have the distinguishing property that each point of a manifold has a neighbourhood diffeomorphic with an open ball of a Euclidean space, and when one deals with manifolds, this fact is used extensively. However, in the equivariant setting, one wants to deal with equivariant neighbourhoods. In this chapter, we construct equivariant substitutes for coordinate neighbourhoods, which are in effect orbits of neighbourhoods diffeomorphic to a Euclidean ball. Furthermore, we will construct the neighbourhoods in such a way that in the coordinates of the Euclidean ball, the action is orthogonal.

We start by studying the twisted product construction. The twisted product is strongly connected to our substitute for coordinate neighbourhoods, since our substitute for a coordinate neighbourhood turns out to be the twisted product of the acting group G and the Euclidean ball.

1.1 Twisted product

In this section, we state the basic properties of the twisted product, see Kawakubo [13], Section 2.2 for further information. Furthermore, we prove some useful lemmas about the twisted product in the situation where maps are C^r and/or the acting group has the discrete topology.

Let $H < G$, where G is a group. (In the general case G is a topological group, but since we are interested in properly discontinuous actions, it is a discrete group.) Let H act on a space X. In this section, unless stated otherwise, we do not assume that the action is properly discontinuous.

The twisted product $G \times_H X$ is constructed as follows: We start with the space $G \times X$ and consider there the action of H defined by $h(g, x) = (gh^{-1}, hx)$ for all $h \in H$. Then $G \times_H X = (G \times X)/H$. We denote the equivalence class of (g, x) in $G \times_H X$ by $[g, x]$. There is an action of G in $G \times_H X$ such that $g[g', x] = [gg', x]$ for all $g \in G$.

In the following, we will assume that G is a discrete group. In this case, the twisted product is quite simple.

Lemma 6 *The space $G \times_H X$ is a disjoint union of G/H copies of X such that all the copies are open in $G \times_H X$. Furthermore, the subset $[e \times X] \subset G \times_H X$ is H-equivariantly homeomorphic to X, and each $g \in G$ takes $[e \times X]$ homeomorphically to some of the G/H copies of X. Thus we identify X with $[e \times X]$.*

Proof: Let G' be a set of representatives for the equivalence relation $g \sim g'$ if and only if $gh = g'$ for some $h \in H$. Denote by π the orbit map $G \times X \to G \times_H X$, and $\pi_g = \pi | g \times X$. Since π is open as an orbit map, each $\operatorname{im} \pi_g = [g \times X]$ is an open set.

Assume $g \sim g'$. Let $[g, x] \in [g \times X]$. Then there exists $h \in H$ such that $[g, x] = [g'h, x] = [g', hx] \in [g' \times X]$. Assume $g \not\sim g'$. Let $[g, x] \in [g \times X]$. Similarly, $[g', x] \in [g \times G]$, and thus $[g \times X] = [g' \times X]$. Then for all $h \in H$ we have that $g'h \neq g$, and, consequently, for all $y \in X$ we have $[g, x] \neq [g', y]$. Thus we have the following: If $g \sim g'$, then $[g \times X] = [g' \times X]$, and if $g \not\sim g'$, then $[g \times X]$ and $[g' \times X]$ are disjoint. Thus $G \times_H X$ is the disjoint union of the subspaces $[g' \times X]$, $g' \in G'$.

Assume $\pi_g(g, x) = \pi_g(g, y)$. Then $(g, x) = (gh, h^{-1}y)$ and thus $h = e$ and $x = y$. Thus π_g is injective, and since the orbit map is open, we have that π_g is a topological imbedding of $g \times X$ onto $[g \times X]$ for each $g \in G$, in particular, for $g = e$. If $h \in H$, then $\pi_e(e, hx) = [e, hx] = [h, x] = h[e, x] = h\pi_e(e, x)$, and thus π_e is H-equivariant.

We have so far the following: The space $G \times_H X$ is a disjoint union of the subspaces $[g' \times_H X]$, $g' \in G'$, and each $\pi_g \colon g \times X \to [g' \times X]$ is a homeomorphism, where $g' \in G'$ is the unique element such that $g' \sim g$. Moreover, π_e is H-equivariant.

To complete the proof, we need to show that each $g \in G$ is such that $g\pi_e \colon [e \times X] \to [g' \times X]$ is a homeomorphism, where $g' \sim g$. But we have that $g\pi_e(e, x) = g[e, x] = [g, x] = \pi_g(g, x)$, and thus $g \circ \pi_e = (\pi_g \circ g) | e \times X$, which implies that $g | [e \times X] = (\pi_g \circ g \circ \pi_e^{-1}) | [e \times X]$, which is a homeomorphism onto $[g' \times X]$, where $g' \sim g$. The proof is complete.□

If, furthermore, X is a C^r-manifold and all $h \in H$ are C^r diffeomorpmisms, then $G \times_H X$ has a natural C^r-structure such that all $g \in G$ are C^r.

The structure is constructed as follows: We start with $G \times X$ and determine a C^r structure for each $g \times X$ by declaring each of the maps $i_g \colon X \to g \times X$, $i_g(x) = (g, x)$ to be diffeomorphisms. Let G' and $\pi_{g'}$ be as in the proof of Lemma 6. We know that each of the maps $\pi_{g'}$, $g' \in G'$ is a topological embedding onto an open subset of $G \times_H X$, and if $g' \neq g'' \in G'$, these subsets are disjoint. We determine the C^r-structure on $G \times_H X$ so that the maps $\pi_{g'}$, $g' \in G'$ are diffeomorpmisms.

Next we check that the construction above works indeed.

Lemma 7 *The structure determined above is independent of the choice of G'.*

Proof: Each of the maps $a_h = h\colon G \times X \to G \times X; h(g,x) = (gh, h^{-1}x)$ is a C^r-diffeomorpmism.

Assume π_g, $\pi_{g'}$ are maps onto $[g \times X]$. Then $g' = gh$, and $\pi_{g'} = \pi_g a_h$. Since a_h is a diffeomorphism, we have that π_g' is a diffeomorphism if and only if π_g is a diffeomorphism.□

Lemma 8 *The action in $G \times_H X$ defined by $g[g', x] = [gg', x]$ is an action by diffeomorphisms.*

Proof: Since $G \times_H X$ is a disjoint union of sets of the type $[g \times X]$ (where g runs through a suitable set of representatives), it is enough to prove that each of the maps $g'|[g \times X]$ is a diffeomorphism. But this is immediate from the decomposition $g'|[g \times X] = \pi_{g'g} \circ g'|(g \times X) \circ \pi_g^{-1}$, where all maps on the right-hand side of the equation are diffeomorphisms. □

Thus, we have a good differentiable structure in $G \times_H X$.

If Y is a G-space, and $f\colon X \to Y$ is an H-equivariant map, then there is a unique map $G \times_H f\colon G \times_H X \to Y$ given by $[g,x] \mapsto gf(x)$ such that $G \times_H f$ is G-equivariant and $(G \times_H f)|X = f$. We will be using this fact extensively, when we define a map f in $e \times X$ and want to extend it to the entire $G \times_H X$, and may occasionally call this process *copying a map f with the action of G.*

Kawakubo [13], Section 2.2, contains the proof of the existence of $G \times_H f$ only in the case where $Y = G \times_H Y'$. Thus we show the existence in our full case.

Lemma 9 *Let X be an H-space and Y be a G-space. Let $f\colon X \to Y$ be H-equivariant. Then there exists a unique $G \times_H f\colon G \times_H X \to Y$ extending f.*

Proof: We prove the existence. Let $G \times f\colon G \times X \to Y$ be the map defined by $(g,x) \mapsto gf(x)$. Then f is continuous, since it is continuous in each $g \times X$, which are open.

Let $(g,x), (g',x') \in G \times X$ such that $[g,x] = [g',x']$. Then there exists $h \in H$ such that $g' = gh$ and $x' = h^{-1}x$. Then, $G \times f(g',x') = g'f(x') = ghf(h^{-1}x) = ghh^{-1}f(x) = gf(x) = G \times f(g,x)$. Thus, $G \times f$ factorizes through $G \times_H X$ as $G \times f = (G \times_H f) \circ \pi$. Since $G \times f$ is continuous and π is a quotient map, $G \times_H f$ is continuous. $G \times_H f$ is is equivariant, since $G \times_H f(g[g'x]) = G \times_H f[gg',x] = gg'f(x) = g(G \times_H f[g',x])$.

We prove the uniqueness. Let f' be another G-equivariant map $G \times_H X \to Y$ such that $f'|X = f$. Then, $f'[g,x] = gf'[e,x] = gf(x) = G \times_H f[g,x]$. Thus, $f' = G \times_H f$. □

Also here differentiability produces no problems:

Lemma 10 *If f is a C^r-map, then also $G \times_H f$ is a C^r-map.*

Proof: It is enough to prove that $G \times_H f|[g \times X]$ is a C^r map for each g, since the sets $[g \times_H X]$ are open. But $G \times_H f|[g \times X] = g \circ f \circ g^{-1}$, which is a C^r-map.□

We will identify a twisted product with any space that is equivariantly homeomorphic (diffeomorphic) with it.

When the acting group is discrete, the twisted product has (up to the equivariant isomorphism) a relatively nice characterization, which we establish next and derive some consequences of this characterization.

Lemma 11 *Let $G \curvearrowright X$, where G is discrete, and $H < G$. Assume $A \subset X$ satisfies the following conditions:*

1. *A is open.*

2. *$GA = X$.*

3. *For all $a \in A$, we have that $ga \in A$ if and only if $g \in H$.*

Then $G \times_H i \colon G \times_H A \to X$ is an isomorphism, where $i \colon A \hookrightarrow X$.

If, furthermore, X has a C^r differentiable structure, and all $g \in G$ are C^r-differentiable, then the identification map $G \times_H i \colon G \times_H A \to X$ is a diffeomorphism.

Proof: We have the map $G \times_H i \colon G \times_H A \to X$, where i is the canonical H-equivariant inclusion. We need to check that it is a homeomorphism. It is surjective by (2), and we need to check that it is injective. Assume $[g, a], [g', a'] \in G \times_H A$ such that $(G \times_H i)[g, a] = (G \times_H i)[g', a']$. Now $g'a' = ga$, which implies $g^{-1}g'a' = a$, and thus $g^{-1}g' \in H$. Thus in the twisted product, $[g, a] = [g, g^{-1}g'a'] = [gg^{-1}g', a'] = [g', a']$.

That $G \times_H i$ is homeomorphism follows, since each $[g, a] \in G \times_H A$ has a neighbourhood $g \times_H A$ such that its image is gA, which is open in X, since A is open in X. Furthermore, one sees that $(G \times_H i)|(g \times_H A) = g \circ i \circ g^{-1}|(g \times_H A)$ is a topological embedding as a composition of embeddings.

We prove that $G \times_H i$ is a diffeomorphism. It is enough to prove that for each gA, $(G \times_H i)|gA$ is C^r and an immersion. But now $(G \times_H i)|gA = g \circ i \circ g^{-1}$, and all the maps g^{-1}, i and g are C^r and immersions. □

The rest of this section consists of corollaries to the previous lemma. We will assume without further notification that G is a discrete group.

Corollary 12 *Let X be a G-space, G discrete, $H < G$. Let $A \subset X$ be a H-invariant subset, i the canonical embedding. Then $G \times_H i$ is an isomorphism if and only if A and X satisfy the numbered assumptions of the previous lemma.*

Proof: The previous lemma is the *if* direction. To prove the *only if* direction, it is enough to check that $G \times_H A$ satisfies the assumptions of the previous lemma. Now $G \times_H A$ is the orbit space of $G \times A$ under the action of H. Let $a \in A$. Then $[g, a] \in A$ if and only if there exists $h \in H$ such that $gh = e$, and since H is a subgroup, this is true if and only if $g = h^{-1} \in H$. This gives us (3). To prove (1), we remark that the orbit map is always open, and $g \times A$ is open in $G \times A$. Thus $G \times_H A$ satisfies (1). Moreover, in $G \times A$ we have that $G(e \times A) = G \times A$, and by passing to the quotient, we see that $GA = G \times_H A$. Thus we have (2). □

Corollary 13 *Let X be a G-space, and A an open connected component (e.g. any connected component in a locally connected space) of X such that $GA = X$. Let $H < G$ be the subgroup of those $g \in G$ for which $gA = A$. Then $X = G \times_H A$.*

Proof: We have to check that H satisfies (3) of Lemma 11. But this is obvious, since each $g \in G$, as a homeomorphism, maps a connected component of X onto a connected component of X. Thus, either $gA = A$, in which case $g \in H$, or gA is a connected component of X different from A, and $gA \cap A = \emptyset$.□

Corollary 14 *If $Y = G \times_H B$ and $f\colon X \to Y$ is G-equivariant, then $X = G \times_H f^{-1}B$.*

In particular, if $Y = G \times_H B$, and $A \subset B$ is H-invariant, then $G \times_H A = GA \subset Y$.

Proof: We know that B satisfies the assumptions of Lemma 11. We prove that $f^{-1}B$ satisfies them, too. Since B is open in Y and f is continuous, $f^{-1}B$ is open in X. This is (1). Let $x \in X$. Then there exists $g \in G$ such that $gf(x) \in B$. But now $gx \in f^{-1}B$. This is (2). Let $g \in G$ and $b \in f^{-1}B$. Now $gf(b) \in B$ if and only in $g \in H$. Thus $f(gb) \in B$ if and only if $g \in H$. Thus $gb \in f^{-1}B$ is and only if $g \in H$. This is (3).

The *in particular* follows by applying the first part to the canonical inclusion $GA \hookrightarrow Y$.□

Corollary 15 *If G is a discrete group and $H < G$ acts properly discontinuously on X, then G acts properly discontinuously on $G \times_H X$.*

In particular, if H is finite, then G acts properly discontinuously on $G \times_H X$.

Proof: Let $x, y \in G \times_H X$. Let $g \in G$ such that $gx \in X$ and let $g' \in G$ such that $g'y \in X$. Let $U \subset X$ be a neighbourhood of gx and $V \subset X$ be a neighbourhood of gy such that $U \cap hV \neq \emptyset$ only for finitely many $h \in H$. Because of (3) in Lemma 11, we have that $U \cap g''V = \emptyset$ if $g'' \in G \setminus H$. Thus $U \cap g''V \neq \emptyset$ only for finitely many $g'' \in G$. Let G'' be the set of such g''.

Since X is open in $G \times_H X$, we have that U, V are open in $G \times_H X$, and, consequently, $U' = g^{-1}U, V' = g'^{-1}V$ are neighbourhoods of x, y, respectively. Furthermore, $U' \cap g'''V' \neq \emptyset$ if and only if $gU' \cap gg'''V' \neq \emptyset$ if and only if $U \cap gg'''g'^{-1}V \neq \emptyset$ if and only if $gg'''g'^{-1} \in G''$, which is true only for finitely many $g''' \in G$. Thus the action is properly discontinuous.

The *in particular* part follows, as one observes that an action of a finite group is always properly discontinuous.□

Now we have a good and easy-to-check characterization for twisted products, namely, the three numbered conditions of Lemma 11. Since twisted products have the map extension property, we have from now on the following technique in use: If X, Y are a G-spaces and $A \subset X$ satisfied the assumptions of Lemma 11, then each H-equivariant $f\colon A \to Y$ extends to a G-equivariant $G \times_H f\colon X \to Y$.

Before proceeding, we check the rather trivial fact that this extension technique preserves injectivity.

Lemma 16 *Let X be a G-space and $H < G$. Let X' be an H-invariant subspace of X such that $X = G \times_H X'$. If A is a H-space and $f\colon A \to X'$ is an H-equivariant injective map, then*

$$G \times_H f\colon G \times_H A \to X$$

is injective.

Proof: Let $[g, x], [g', x'] \in G \times_H A$ be such that $(G \times_H f)[g, x] = (G \times_H f)[g', x']$. Then, by the definition of $G \times_H f$, we have that $gf(x) = g'f(x')$, and thus ${g'}^{-1}gf(x) = f(x')$. Since both $f(x)$ and $f(x')$ are in X' and $X = G \times_H X'$, this implies that ${g'}^{-1}g \in H$. Since f is H-equivariant, we have $f({g'}^{-1}gx) = f(x')$, and since f is injective, ${g'}^{-1}gx = x'$. Thus $[e, {g'}^{-1}gx] = [e, x']$, which implies $[g, x] = [g', x']$.□

Acknowledgements

Lemma 11 and Corollary 13 were invented by me, and I do not have any knowledge of their prior instances (except for my Licentiate Thesis). Everything else is either well-known or trivial consequences of well-known facts. However, it was easier to write everything down with proofs than try to find a source to cite.

The twisted product in the C^∞- and C^ω-categories can be found, for example, in Illman [8].

1.2 Coordinate neighbourhoods

Here we prove that each point of a C^r G-manifold pair (M, N) has a coordinate neighbourhood such that the coordinate structure of the neighbourhood is compatible with the action. Here, *coordinate neighbourhood* must be understood in a generalized sense; see the following definition.

Definition 17 *We say that a G-space X is a generalized coordinate space (or, briefly, a coordinate space) if X is topologically a disjoint union of copies of $\mathbb{R}^n$ for a fixed n, the action is properly discontinuous and C^r, $X = GU$ and G_U is finite for some (equivalently: all) components U of X.*

Furthermore, we assume that there is a special component X' of X such that the action of $G_{X'}$ on X' is orthogonal when X' has a fixed coordinate structure of $\{x \in \mathbb{R}^n \mid |x| < \epsilon\}$ for some $\epsilon > 0$. By the coordinate structure of $\{x \in \mathbb{R}^n \mid |x| < \epsilon\}$ we mean a diffeomorphism $\{x \in \mathbb{R}^n \mid |x| < \epsilon\} \to X'$, which we use to determine local coordinates.

We make here the convention that whenever we have a coordinate space X, we denote its special component with the coordinate structure by X'.

Definition 18 *We say that (X, X_0) is a generalized coordinate space pair, if X_0 and X are coordinate spaces and $X_0 \subset X$. We further require that $X' \cap X_0 = X_0'$, and $X_0' = \{x \in X' | x = (x_0, \dots, x_{n-1}, 0, \dots, 0), x_0, \dots x_{n-1} \in \mathbb{R}\}$, where $n = \dim X_0$.*

Definition 19 *We say that a G-space X is a generalized coordinate half-space if X is a disjoint union of copies of H^n for a fixed n, the action is properly discontinuous and C^r differentiable, $X = GU$ and G_U is finite for some (equivalently: all) components U of X.*

Furthermore, we assume that there is a special component X' of X such that the action of $G_{X'}$ on X' is orthogonal when X' has a fixed structure of $\{x \in \mathbb{R}^n \mid |x| < \epsilon, x_0 \geq 0\}$ for some $\epsilon > 0$.

A coordinate half-space pair is defined similarly, so that it is a pair of coordinate half-spaces.

Definition 20 *Let (M, N) be a C^r G-manifold pair and $x \in (M, N)$. We say that U is a coordinate neighbourhood of x if $x \in U$, U is open in M, each component of U is relatively compact, the closures of different components of U do not intersect, and one of the following holds:*

- *$x \in M \setminus \partial M$, $x \notin N$, and U is equivariantly C^r-diffeomorphic to a generalized coordinate space such that $U \cap N = \emptyset$.*
- *$x \in \partial M$, $x \notin N$, and U is equivariantly C^r-diffeomorphic to a generalized coordinate half-space such that $U \cap N = \emptyset$.*
- *$x \in M \setminus \partial M$, $x \in N$, and U is equivariantly C^r-diffeomorphic to a generalized coordinate space pair (X, X_0) such that $N \cap X = X_0$.*
- *$x \in \partial M$, $x \in \partial N$, and U is equivariantly C^r-diffeomorphic to a generalized coordinate half-space pair (X, X_0) such that $N \cap X = X_0$.*

We further require that the diffeomorphisms above map x to 0 of the special component having the linear structure.

In all the cases mentioned, we will identify U with the coordinate space, and, in particular, assume that one component of U has a fixed equivariant linear structure.

Remark 21 *In the above definition we have $G_{X'} = G_x$. This is true, since $x = 0$ in the coordinates of X', and each $g \in G_{X'}$ is linear, and thus keeps 0 fixed.*

Next, we establish the connection between coordinate neighbourhoods and twisted products and show how this connection can be used to define equivariant maps, which will provide to be a very important technique when triangulating manifolds.

Lemma 22 *Let X be a coordinate space or a half-space. We have that $X = G \times_{G_{X'}} X'$.*

Proof: Since X' is a connected component of X, the lemma follows by Corollary 13. □

Thus, by Lemma 10 and the discussion preceeding it, we can extend a $G_{X'}$-equivariant map $X' \to M$, where M is a G-space, uniquely to a G-equivariant map $X \to M$.

Remark 23 *We can use the following technique when constructing maps.*

Let $f\colon M \to M'$ be G-equivariant and X a coordinate neighbourhood in M. If we define $f''\colon X' \to M'$ such that f'' is $G_{X'}$-equivariant and $\{x \mid f''(x) \neq f(x)\} \subset\subset X'$, then we can extend f'' by the twisted product construction to $G \times_{G_{X'}} f'' = f'|X\colon X \to M'$ and define $f'\colon M \to M'$ as follows: $f'(x) = (f'|X)(x)$ if $x \in X$, and $f'(x) = f(x)$ otherwise. Now, obviously, f' is G-equivariant, and it is continuous (differentiable) if f and $f'|X'$ are. The last fact is true, since each $x \in M$ has a neighbourhood V_x such that $f'|V_x$ equals either $f''|V_x$ or $f|V_x$.

We call this operation changing f by f'' and copying changes with the action with f' as the result.

The same technique can also be used in the following situation. Let $f\colon M \to M'$ be as before, but $X \subset M'$. Then we have that $f^{-1}X = G \times_{G_{X'}} f^{-1}X'$ by Corollary 14 and the fact that $X = G \times_{G_{X'}} X'$.

Now "changing an copying" works in exactly the same way, if $f''|f^{-1}X'$ is $G_{X'}$-equivariant and $\{x \mid f''(x) \neq f(x)\} \subset\subset f^{-1}X'$.

We use the coordinate structure of X' to take equivariant linear combinations. We prove that it is possible.

Lemma 24 *Let M be a space with an action of G, X a coordinate space, and let f, h be G-equivariant maps $M \to X$ such that $f^{-1}X' = h^{-1}X'$. Then, for each $t \in [0,1]$, the map $tf + (1-t)h$ is well-defined and G-equivariant.*

Proof: The special component X' is convex in its Euclidean structure. Thus $(tf + (1-t)h)|f^{-1}X'$ is well-defined. It is $G_{X'}$-equivariant, since

$$tf(gx) + (1-t)h(gx) = tgf(x) + (1-t)gh(x) = g(tf(x) + (1-t)h(x)).$$

The last equality is true, since g is linear.

By Lemma 14, $M = G \times_{G_{X'}} f^{-1}X'$. Thus, $(tf + (1-t)h)|X'$ extends uniquely into an equivariant map $M \to X$, and we declare that this is the map $tf + (1-t)h$.□

Next, we prove by a sequence of lemmas that each $x \in (M, N)$ has a coordinate neighbourhood.

First, we do not assume that M has an action of G, and thus we use the expression *chart* to mean a local coordinate system which does not have to be compatible with the action.

The following lemma is well-known in the literature, but we include it for the sake of completeness.

Lemma 25 *Let M be a manifold, possibly with a boundary, and N a submanifold of M. Let x be a point in N. If $x \in \partial N$, we, furthermore, assume that $x \in \partial M$ and T_xN is transversal to $T_x\partial M$. Then there exists a chart U in x such that $N \cap U$ is a chart in N (i.e., $N \cap U$ is a linear subspace in the coordinates of U.)*

Proof: Let the dimension of M be n and the dimension of N be p.

Assume first that $x \notin \partial M$, $x \notin \partial N$. Then start by restricting the attention into a small enough chart V about x such that it does not meet the boundary and work exclusively inside it. Choose coordinates in V so that x lies at the origin. Choose a coordinate map $w\colon W \to N \cap V$ at x so that x is the origin. In the linear space T_xV choose a basis $e = (e_1, \dots e_p, e_{p+1}, \dots, e_n)$ so that the embedding $D_{w^{-1}x}w$ is of the form $(x_1, \dots, x_p) \mapsto (x_1, \dots, x_p, 0, \dots, 0)$ in the basis e. Since V is homeomorphic to an open set of $\mathbb{R}^n$, we can assume that all e_i are vectors of $\mathbb{R}^n$, and can be naturally added to points of V. Define $f\colon X \subset \mathbb{R}^n \to V$ to be the map $f(x_1, \dots, x_n) = w(x_1, x_2, \dots, x_p) + x_{p+1}e_{p+1} + \dots + x_ne_n$, where the addition is done by using the chosen coordinates of V, and $X \subset \mathbb{R}^n$ is a small neighbourhood of the origin. Then the tangent map of f is a bijection at the origin. Thus, f is an immersion at the origin, and the proof of the current case is complete, as we choose as U a small enough subset of V. Then, by Munkres [18], Corollary 1.12, f is a diffeomorphism in $f^{-1}U$ for a small enough U.

Assume then that $x \in \partial M$, $x \in \partial N$ and T_xN is transversal to $T_x\partial M$. Proceed as before, but when choosing the basis $e_1, e_2, \dots,$ make the choices in such a way that e_1 does not lie inside $T_x\partial M$ and all the other e_i's do. Since we have made the assumption about transversality, w, as a coordinate map, is automatically of this form. Since the dimension of the boundary is exactly one smaller than the dimension of the manifold, it is possible to choose the rest of e_i's in the required fashion. Since V is a chart such that ∂M coincides with the hyperplane $\{(0, y_2, \dots, y_n) \mid (y_2, \dots, y_n) \in U' \subset \mathbb{R}^{n-1}\}$, we have (for small enough $(y_1, \dots, y_n)$) that $f(y_1, \dots, y_n) \in \partial M$ if and only if $y_1 = 0$, thus f is a coordinate map.□

Lemma 26 *Assume that (X, A) and (Y, B) are topological spaces and let $x \in A$. Assume that $f\colon (X, A) \to (Y, B)$ is a continuous injection such that $f(A)$ is a neighbourhood of $f(x)$ in the space B ($f(A)$ not necessarily open in the space B). Then there exists a neighbourhood U of x such that $f(A \cap U) = B \cap f(U)$.*

Proof: Let V' be a neighbourhood of $f(x)$ open in the space B such that $V' \subset fA$, and let V be an open subset of Y such that $V \cap B = V'$. Let U be a neighbourhood of x such that $fU \subset V$. We claim that U is the desired neighbourhood. If not, then there exists $y \in U \setminus A$ such that $f(y) \in B$. But now $f(y) \in V' \subset fA$, and thus $f(z) = f(y)$ for some $z \in A$ violating the fact that f is injective.□

The following lemma is known for Euclidean spaces, and we check that it holds for half-spaces also.

Lemma 27 *Let $U \subset H^n$ be an open neighbourhood of 0. Let $f\colon U \to H^n$ be a C^r-map such that f is an immersion at 0, $f(0) = 0$ and $f(U \cap \mathbb{R}^{n-1}) \subset \mathbb{R}^{n-1}$. Then there exists an open neighbourhood $V \subset U$ of 0 such that $f|V$ is a diffeomorphism onto an open subset of H^n.*

Proof: Since the partial derivatives of f are continuous, by restricting U we can assume that f is an immersion in U.

Write $f(x) = l_f(x) + e(x)$, where l_f is linear, and the partial derivatives of e are zero at 0. Thus, l_f is Bilipschitz with a Bilipschitz constant c, and in a small enough neighbourhood of 0, the map e is Lipschitz with an arbitrarily small Lipschitz constant d. By choosing d smaller than $1/c$, we see that f is Bilipschitz, and hence injective in a small enough neighbourhood of 0. Thus we can assume that f is injective by shrinking U.

We prove that $\operatorname{im} f$ is open. By the previous lemma and the fact that $f|\mathbb{R}^{n-1}$ is a diffeomorphism in a neighbourhood of 0 onto a neighbourhood of 0 in $\mathbb{R}^{n-1}$, there exists an open neighbourhood V of 0 in H^n such that $f(\mathbb{R}^{n-1} \cap V) = \mathbb{R}^{n-1} \cap f(V)$. Let $V' = \{(x_1, \ldots, x_n) \in \mathbb{R}^n \mid (|x_1|, x_2, \ldots, x_n) \in V\}$. Let $m\colon \mathbb{R}^n \to \mathbb{R}^n$ be the orthogonal reflection with respect to $\mathbb{R}^{m-1}$. Let $f''\colon V' \to \mathbb{R}^n$ be the map $f''(x) = f(x)$, if $x \in H^n$, and $f''(x) = mfm(x)$, if $x \notin H^n$. Now f'' is continuous and injective (since $f(\mathbb{R}^{m-1} \cap V) = \mathbb{R}^{n-1} \cap f(V)$, and by invariance of domain, $\operatorname{im} f'$ is open. But $\operatorname{im} f = H^n \cap \operatorname{im} f'$, that is, open in H^n.□

Lemma 28 *Let $U \subset \mathbb{R}^n$ or $U \subset H^n$ be an open neigbourhood of 0, and assume that there exists a C^r-action of a finite group H on U such that 0 is a fixed point of H.*

Then there exists neighbourhoods $V, V' \subset U$ of 0 and a C^r-diffeomorphism $T\colon V \to V'$ such that $T(0) = 0$ and each of the maps ThT^{-1} is a restriction of a linear map.

Furthermore, for each linear subspace L of $\mathbb{R}^n$ such that $L \cap U$ is H-invariant, there exists a H-invariant neighbourhood V'' of 0 such that $T(V'' \cap L) = T(V'') \cap L$.

Proof: (Based on the proof given in Bochner [2], Theorem 1.)

Represent each map $h \in H$ in the form $h(y) = l_h(y) + \epsilon_h(y)$, where l_h is a linear map, and $\epsilon_h(y) = |y|e_h(y)$ for some map e_h converging to zero when y tends to 0.

For each h, define map $f_h = (l_h)^{-1} \circ h$. Now, f_h is of the form $\mathrm{id} + \epsilon'$, where ϵ' is of the form $|y|e'_h(y)$, where e'_h converges to zero when $y \to 0$.

(If the ambient space is H^n, we remark the following: Since the action of H restricts to a C^r-action of $U \cap \mathbb{R}^{n-1}$, we have that $f_h U \cap \mathbb{R}^{n-1} \subset \mathbb{R}^{n-1}$. Also, $f_h U \subset H^n$ for each $h \in H$.)

Define $T = \frac{\sum_{h\in H} f_h}{|H|}$ (the finite version of the Haar integral). Now T is of the form $\mathrm{id} + \epsilon''$, that is, it is immersion at 0. (If the ambient space is H^n, $T(\mathbb{R}^{n-1} \cap U) \subset \mathbb{R}^{n-1}$, and $TU \subset H^n$, since T is an average of maps f_h satisfying these conditions.)

Thus T is a diffeomorphism in some neighbourhood of 0 by Munkres [18], Corollary 1.12, if the ambient space is $\mathbb{R}^n$, and by Lemma 27, if the ambient space is H^n.

Now let $k \in H$. We have that

$$l_k \circ T = \frac{\sum_{h\in H} l_k \circ l_h^{-1} \circ h}{|H|} = \frac{\sum_{h\in H} l_{kh^{-1}} \circ hk^{-1}k}{|H|} =$$

$$\frac{\sum_{h\in H} l_{hk^{-1}}^{-1} \circ hk^{-1}}{|H|} k = \frac{\sum_{h\in H} f_{fk^{-1}}}{|H|} k = T \circ k.$$

Thus $l_k = TkT^{-1}$, that is, a linear map.

Assume now that V is an invariant neighbourhood of 0 such that $T|V$ is a diffeomorphism $V \to V'$.

Assume now that L is a linear subspace of $\mathbb{R}^n$ such that $L \cap U$ is H-invariant. We easily see that each l_h and h map a point of $L \cap U$ into a point of L. Thus T, composed of the maps l_h and h, $h \in H$, maps a point of $L \cap V$ into a point of L. Thus $T(L \cap V) \subset L$.

Now the existence of V'' follows from Lemma 26 applied to the map $T\colon (V, V \cap L) \to (V', V' \cap L)$. The set $T(V \cap L)$ is a neighbourhood of 0 in $V' \cap L$, since $\dim(V \cap L) = \dim(V' \cap L)$, T is an immersion, and thus $T|\colon (V \cap L) \to (V' \cap L)$ restricts to a diffeomorphism in some neighbourhood of 0. By shrinking V'' we can assume that it is H-invariant. □

Corollary 29 *Let M be a G-manifold, $H < G$, and $x \in M$ such that $G_x = H$. By choosing a suitable $V \subset \mathbb{R}^n$ (or $V \subset H^n$), we can define a H-equivariant coordinate map $T^{-1}\colon V \to U \subset M$ when we define $H \curvearrowright V$ by $h(y) = ThT^{-1}(y) = l_h(y)$.*

Proof: If U is a chart in M, we can identify U with a subset of $\mathbb{R}^n$ (or H^n, if $x \in \partial M$) giving the coordinates. By Lemma 4, we can shrink V and V' given by the previous lemma such that they are H-invariant.□

Lemma 30 *If, M is a G-manifold, x is a boundary point of M, $G_x = H$, and V is an H-equivariant coordinate neighbourhood of x with coordinates where the action of H is linear, $y = (y_1, \ldots, y_n) \in V$ and $h \in H$, then the first (i.e., the one transversal to the boundary) coordinate of $h(y_1, \ldots, y_n)$ must be y_1.*

Proof: If the first coordinate of $h(y_1, \ldots, y_n)$ depends on some other y_i than y_1, then, by choosing all the other coefficients to be zero and the coefficient y_i non-zero, one goes to the wrong side of the half-plane by h in every neighbourhood of x. A contradiction. If the coefficient of the first coordinate in the linear combination representation of $h(y_1, 0, \ldots, 0)$ is smaller than y_1, say λy_1, $0 < \lambda < 1$, then the coefficient of the first coordinate of $h^n(y_1, 0, \ldots, 0)$, $n \in \mathbb{Z}_+$ is less than $\lambda^n y_1$, and the orbit of y_1 converges to the boundary. A contradiction with finiteness of H_x. If it is $\lambda > 1$, one obtains a similar contradiction by considering h^{-1}.□

Proposition 31 *With the restriction about the transversality stated above, each point in a manifold pair (M, N), with a properly discontinuous action of G, has a generalized coordinate neighbourhood.*

(Here we assume just that it is enough to have a linear action in the special component. We will obtain an orthogonal action in the next corollary.)

Proof: Let $x \in M$. Since the action is properly discontinuos, $\min_{g \in G, x \neq gx} d(x, gx)$ is positive, let it be $3d$. Now the neighbourhood $B(x, d)$ contains a chart. If $x \in M \setminus N$, the neighbourhood can still be diminished to obtain a chart that does not meet N, and if $x \in N$, we can use Lemma 25 to obtain a coordinate structure that is compatible with the pair structure. By Lemma 4, we can assume that the chart is G_x-invariant.

Next, we get the action of G_x linear in suitable coordinates by applying Lemma 28 and Corollary 29. By the furthermore-part in Lemma 28, if $x \in N$, the resulting coordinate structure can be made into a coordinate structure of pairs by shrinking the chart, that is, the coordinates of N are of the type $(x_1, \ldots, x_n, 0, \ldots, 0)$. Let now $f \colon U \to M$ be the coordinate map we have obtained so far. We finally get the G-invariant neighbourhood by the twisted product construction $G \times_{G_x} f \colon G \times_{G_x} f \to M$. We remark that the choice of d together with the invariance of the metric ensure that $\bar{U}$ and $g\bar{U}$ intersect only if $g \in G_x$. □

Corollary 32 *In the generalized coordinate neighbourhoods X, the special component X' can be equipped with a $G_{X'}$-invariant metric so that X' is differentiably isometric with some $B(0, \epsilon)$ (diminish X if necessary), and that the isometry preserves the linear structure given by Lemma 28, and the action is orthogonal in the new coordinates of $B(0, \epsilon)$.*

Furthermore, the diameter of the special component can be chosen arbitrarily small in any chosen metric of the whole manifold M.

Proof: Let X' be equipped with coordinates such that all $g \in G_{X'}$ are linear. Now X' is an open subset of some $\mathbb{R}^n$, and all $g \in G_{X'}$ are linear. Now, by Bredon [3], Theorem 3.5, $\mathbb{R}^n$ has an invariant inner product, which gives us the desired metric and the orthogonality. Now, if necessary, replace X' with a smaller $B(0,\epsilon)$.

The last part of the corollary follows, since any two metrics (the chosen one in the manifold and the Euclidean metric in $B(0,\epsilon)$) are uniformly equivalent in the compact set $\bar{B}(0,\epsilon/2)$, and one can choose $B(0,\epsilon')$ with an arbitrarily small ϵ'. □

Remark 33 *Let d be invariant Euclidean distance in X' given by Corollary 32. We may occasionally call $d(0,x)$ the norm of x.*

Corollary 34 *If $x \in \partial M$ and X is a coordinate neighbourhood of x, then the coordinates in X' can be chosen so that G_x acts on $X' \subset H^n = \mathbb{R}_+ \oplus \mathbb{R}^{n-1}$ in the $n-1$ last coordinates.*

Proof: Take coordinates so that they are given by an orthogonal basis, orthogonal in the sense of the inner product of the previous corollary, and all but one of the basis elements lie in last summand of $\mathbb{R}_+ \oplus \mathbb{R}^{n-1}$. Then, since the action preserves the inner product and ∂M, there is by Corollary 30 exactly one way to choose $g(r,0,\ldots,0)$ for $(r,0,\ldots,0) \in X'$ so that it is orthogonal to ∂M, and the resulting action is properly discontinuous, namely $g(r,0,\ldots,0) = (r,0,\ldots,0)$. Now the claim follows from the linearity of the action in the chosen coordinates. □

Acknowledgements

Our concept of a coordinate neighbourhood, together with its representation as a twisted product, is a generalization of a similar concept in Illman, [7] for a finite G. Apart from a properly discontinuous G, we also improved Illman's argument in another place. Illman uses the tubular neighbourhood theorem to obtain the linear equivariant structure, which works well for C^∞- and C^ω-cases. However, if the original manifold is C^r, $r < \infty$, this technique would yield only a C^{r-1} chart. By Lemma 28, we can circumvent this restriction.

1.3 Coverings with coordinate neighbourhoods

In the previous section, we proved that each point in a manifold has coordinate neighbourhoods. Here we prove that the entire manifold can be covered with a locally finite collection of coordinate neighbourhoods.

Lemma 35 *Let X be a locally compact, paracompact space, and let $\mathcal{B}$ be a basis for the topology of X. Then, each open covering (U_i) of X has a locally finite refinement with elements from $\mathcal{B}$.*

Proof: Since each open covering of X can be refined by a locally finite open covering with relatively compact sets, we can assume that (U_i) is locally finite, and each U_i is relatively compact.

Let (V_j) be a refinement of (U_i) with elements from $\mathcal{B}$ and let (W_k) be a locally finite refinement of (V_j). For each k, choose a V_k, an element of (V_j), such that $W_k \subset V_k$. We claim that the collection (V_k) of the chosen elements V_k is the required covering.

Since (W_k) is a covering, also (V_k) is a covering. Furthermore, each V_k is an element of $\mathcal{B}$.

We prove that (V_k) is locally finite. Assume first that every U_i contains only finitely many of the sets V_k. Let $x \in X$. Since (U_i) is a locally finite covering, there exists a neighbourhood U of x such that $U \cap U_i \neq \emptyset$ only for finitely many U_i. If $V_k \cap U \neq \emptyset$, then $V_k \subset U_i$ for some U_i intersecting U. Since there is only a finite number of such U_i and each of them contains only finitely many V_k, we have that $U \cap V_k \neq \emptyset$ only for finitely many V_k. Thus, in this case, the covering (V_k) is locally finite.

Then we assume that some of the sets U_i contains infinitely many V_k and prove that this leads to a contradiction. Since U_i contains infinitely many of the elements V_k, it contains infinitely many of the elements W_k. Choose from each of those infinitely many W_k a point w_k. Since each $w_k \in U_i$ and U_i is relatively compact, the collection of elements w_k has a cluster point $w \in \bar{U}_i$. Each neighbourhood of w meets infinitely many of the sets W_k, which is a contradiction since (W_k) is locally finite. Thus the proof is complete.□

Let M be a G-space. We recall that we call a covering (U_i) a G-covering if each U_i is G-invariant.

Lemma 36 *Any G-covering $\mathcal{U}$ of a differentiable G-manifold pair (M, N) can be refined by a locally finite open G-covering consisting of generalized coordinate neighbourhoods.*

Proof: By Abels–Strantzalous, [1], Proposition 3.2.5.D, M/G is paracompact.

By Abels–Strantzalous, [1], 1.7.3., $\pi\colon M \to M/G$ is an open map. Let $x \in M/G$ and let U be a compact neighbourhood of a point $y \in M$ such that $f(y) = x$. Now fU is a compact neighbourhood of X. Thus M/G is locally compact.

We claim that the sets πX, X a coordinate neighbourhood in (M, N), form a basis for the topology of M/G. Let $x \in M/G$ and let U be a neighbourhood of x. Let $y \in M$ such that $f(y) = x$. Now, $\pi^{-1}U$ is an invariant neighbourhood of y, and since the diameter of the special component of a coordinate neighbourhood of y can be chosen arbitrarily small, y has a coordinate neighbourhood X such that $X' \subset \pi^{-1}U$. Since U is invariant and $X = GX'$, we have that $X \subset \pi^{-1}U$. Since π is open, we have that πX is a neighbourhood of x such that $\pi X \subset U$. Thus the sets πX form a basis.

Let $\mathfrak{U}$ be a G-covering of M. Now $\pi\mathfrak{U}$ is an open covering of M/G. By Lemma 35, it has a locally finite refinement $\mathcal{V}$ with the sets of the type πX, X a coordinate neighbourhood in M. Now, $\pi^{-1}\mathcal{V}$ is the required refinement.□

The following lemma is well-known, but we prove it for completeness.

Lemma 37 *Let $\mathfrak{U}$ be a locally finite covering with relatively compact sets. Then each $U \in \mathfrak{U}$ meets only a finite number of other members of $\mathfrak{U}$. (In other words, $\mathfrak{U}$ is star-finite.)*

Proof: Assume, on the contrary, that U meets infinitely many V_i, $i \in \mathbb{N}$. Pick for each i an element $x_i \in U \cap V_i$. Then the collection of x_i's has a cluster point x, and each neighbourhood of x meets infinitely many V_i's. □

Corollary 38 *Let $\mathfrak{U}$ be a locally finite G-covering of a G-manifold M with generalized coordinate neighbourhoods. Then each $U \in \mathfrak{U}$ meets only a finite number of other members of $\mathfrak{U}$.*

Proof: Use the previous lemma for the orbit space M/G and the covering $\pi\mathfrak{U}$.□

Let $\mathfrak{U} = \{U_i\}_{i\in I}$ and $\mathcal{V} = \{V_i\}_{i\in I}$ be coverings of a topological space X. Following Pears [20], we say that the coverings are similar if and only if for every $I' \subset I$ we have that

$$\bigcap_{i\in I'} U_i = \emptyset \Leftrightarrow \bigcap_{i\in I'} V_i = \emptyset.$$

Lemma 39 *Given a locally finite G-covering $\mathfrak{U}$ of a G-manifold pair (M, N) with generalized coordinate neighbourhoods there is a locally finite G-covering $\mathcal{V}$ with coordinate neighbourhoods that is similar to $\mathfrak{U}$ and $V_i \subset\subset U_i$ for all $i \in I$.*

Proof: We can assume without loss of generality that $I = \mathbb{N}$. For, if M is not N_2 (second countable), nevertheless each component of M is N_2, and thus, if M is of the type GC, C a component of M, then each locally finite G-covering of M is countable. The general case is a disjoint union of the special cases of the type $M = GC$, each of which can be handled separately.

We can proceed inductively with respect to i, and at each step shrink U_i into V_i. If the similarity requirement is satisfied at every step i (i.e., the covering $(V_1, \ldots V_i, U_{i+1}, U_{i+2}, \ldots)$ satisfies it) then, due to local finiteness, it is satisfied also after infinitely many steps.

But the inductive step i is easy. We have the restriction of a $G_{U'}$-invariant Euclidean norm in the special component U' of U_i, and we can find $\epsilon <$ *the radius of* U' such that when we replace U' with the set $V' = \{u \in U' \mid \|u\| < \epsilon\}$, and U_i with the set $V' \cup (U \setminus U')$, we still have a covering. This is due to the fact that the set $\{x \in U' \mid x \notin V_k, k < i,\ x \notin U_k, k > i\}$ is compact. By invariance of dimension, V' is open, and by enlargening ϵ if necessary, V' satisfies also all of the required intersection conditions (there is only a finite number of them). Now choose $V_i = GV'$. It is as required, and it inherits the required Euclidean structure from U'. The case of a pair does not produce any extra difficulties.$\square$

Putting the results in this chapter together, we have the following:

Theorem 40 *If (V_j) is an open G-covering of a G-manifold pair (M, N), there exiss arbitrarily long sequences of pairwise similar locally finite coverings $(U_i^{'\cdots'})$ with generalized coordinate neighbourhoods satisfying*

$$(U_i) \subset\subset (U_i') \subset\subset (U_i'') \subset\subset \cdots \subset\subset (U_i''''),$$

where all coverings refine (V_j).

Furthermore, all the coverings $(U_i^{'\cdots'})$ are star-finite.

Proof: The existence of the coarsest refinement follows from Lemma 36, and the existence of the rest of the coverings follows from Lemma 39. Star-finiteness follows from Lemma 38.

Acknowledgements

I have not seen Lemmas 35 and 36 anywhere else (except for my Licentiate thesis. The beginning of Lemma 36, that the orbit space of a locally compact space is locally compact, is of course well-known). The non-equivariant form of Lemma 39 is well-known.

Chapter 2

Approximations

Our strategy of triangulating a manifold is to cover the manifold with coordinate neighbourhoods and then triangulate each of those coordinate neighbourhoods. Then we will make *small changes* to the triangulations of parts so that they fit together and form a triangulation of the entire manifold.

In this chapter, we will first define what we mean by a C^r-map between a simplicial complex and a manifold. After that we will introduce the concept of an approximation that we will use to make *small changes* mentioned in the previous paragraph and establish some results concerning it.

2.1 What we count as a map and as an approximation

Let $1 \leq r \leq \omega$.

Let M be a C^r-differentiable G-manifold and $|K|$ a geometrical realization of a simplicial G-complex. (The action is properly discontinuous, as always in this thesis.) We study G-equivariant maps $f\colon |K| \to M$ such that f is continuous, the restriction of f to each simplex is C^r, and each point in M admits a neighbourhood intersecting images of only finitely many simplices. We call such f a map.

Let $|s|$ be an n-simplex, and assume that it has been embedded linearly in $\mathbb{R}^n$. Slightly modifying Munkres [18], Definitions 8.2 and 8.3, we say that $f\colon |s| \to M$ is C^r [immersion] if the following holds for each coordinate neighbourhood X in M.

> Let Y be a component of X. Now Y is diffeomorphic to a ball B of either $\mathbb{R}^{n'}$ or $H^{n'}$. In the latter case, we replace B with the set $B' = \{(x_1, x_2, \ldots, x_{n'}) \in \mathbb{R}^n \mid (|x_1|, x_2, \ldots, x_{n'}) \in B\}$. In the former case, $B' = B$.
>
> Let $A \subset\subset Y$ be compact.
>
> Now we require that for all A and Y of the type characterized above, $f|(f^{-1}A \cap |s|)$ can be extended C^r-differentiably [as an immersion] to some neighbourhood

U of $f^{-1}A \cap |s|$ in $\mathbb{R}^n$ to obtain a C^r-map [immersion] $f'\colon U \to B'$.

(Note that C^r-differentiability and being an immersion are properties of s and f, i.e., they are independent of the choice of the embedding into $\mathbb{R}^n$.)

We say that $f\colon |K| \to M$ is C^r, if the restriction of f to each simplex is C^r.

Remark 41 *If $f|s| \subset X$, where X is a non-boundary coordinate neighbourhood, the condition above is equivalent to the traditional definition that f can be extended into a C^r map $U \to X$, where we assume that $|s|$ lies in $\mathbb{R}^{\dim s}$ and $U \subset \mathbb{R}^{\dim s}$ is a neighbourhood of $|s|$. For the most of this thesis, we will be using simplices that fit inside some coordinate neighbourhood, and thus our definition will coincide with the traditional definition. In particular, the simplices in the existence result of triangulation, Theorem 110, will be of this type. Anyway, by Lemma 112, we can always obtain a situation where each simplex fits inside some coordinate neighbourhood by subdividing enough.*

We need the tricky definition described above, since for boundary neighbourhoods, the extension "$f'\colon U \to X' \hookrightarrow M$" may actually go outside the coordinate neighbourhood and the manifold.

Still following Munkres [18], we say that f is an immersion, if $f||s|$ is an immersion for each simplex s of K, and furthermore the angle of two simplices is positive for each pair of simplices of K having a common vertex. To be exact, the last condition means the following: Let s, s' be simplices such that $x \in |s| \cap |s'|$. Let r and r' be rays starting from x in $|s|$ and $|s'|$, respectively, such that $r \cap |s'| = \{x\}$ and $r' \cap |s| = \{x\}$. Then the positiveness of the angle of $|s|$ and $|s'|$ means that for all such rays r and r', we have that $D_x fr$ and $D_x fr'$ do not intersect outside $0 \in T_{f(x)}M$. Note that since each simplex of $|K|$ has a natural linear structure, a ray in a simplex $|s|$ gives a corresponding ray in $T_x|s|$.

We say that $f\colon |K| \to M$ is an imbedding, if f is an injective immersion. We remark that from the local finiteness of f assumed above, it follows that each embedding is an embedding also in the stronger sense that it is a closed embedding.

A C^r-triangulation t of M is a surjective C^r-embedding $t\colon |K| \to M$. A C^r-triangulation t of a manifold pair (M, N) is a surjective C^r-embedding $t\colon |K, L| \to (M, N)$ such that also $t||L|\colon |L| \to N$ is surjective.

Remark 42 *Differentiability in a simplicial complex is a somewhat weaker property than differentiability in a manifold. This is because differentiability does not control the behaviour of maps at the boundary of simplices very strictly.*

Consider, for example, a simplicial complex K that consists of two 1-simplices, s, s', that have one vertex v in common. Let M, M' be two copies of the unit interval. Now let $t\colon K \to M$ be a differentiable, simplexwise linear map such that $ts = [0, 1/2]$, $ts' = [1/2, 1]$, *and $t'\colon \to M'$ similarly such that $t's = [0, 1/3]$, $t's' = [1/3, 1]$. Now t and t' are bijective C^ω-immersions by the definition above, but tt'^{-1} is not a diffeomorphism of M' and M.*

Thus the definition of the differentiability of $t\colon K \to M$ permits "folds" at the boundaries of simplices.

Lemma 43 *Let K be a simplicial complex and X, Y C^r-spaces. Let $f\colon |K| \to X$ and $f'\colon X \to Y$ be C^r-maps [immersions]. Then $f' \circ f$ is a C^r-map [immersion].*

Proof: If K is a complex of one simplex, the claim is true, since then $f' \circ f$ is in effect a map between spaces. Since a map is C^r iff its restriction to each simplex is C^r, we have that $f' \circ f$ is a C^r-map.

To complete the proof, assume that f', f are immersions, $x \in |K|$, and r, r' are tangent rays starting at x. Then $d_x fr, d_x fr'$ do not intersect outside the origin. Since $D_{f(x)}f'$ is an injective map and $D_x(f' \circ f) = D_{f(x)}f' \circ d_x f$, $D_x(f' \circ f)r$ and $D_x(f' \circ f)r'$ do not intersect outside the origin.□

Next, we handle the case of coordinate half-spaces. First, we give the following definition.

Definition 44 *Let K be a complex and $f\colon |K| \to \mathbb{R}^n$. We say that f is coplanar if $f^{-1}\mathbb{R}^{n-1}$ is a subcomplex of K, and the following holds: Let s be a simplex of K and $x \in |s|, x + v \in |s|$. If $f(x), D_x f(v) \in \mathbb{R}^{n-1}$, then $f(x + v) \in \mathbb{R}^{n-1}$.*

Remark 45 *Note that in the definition of coplanar, as well as in the definition of an immersion, we check only those tangent vectors that "point inside simplices". Although the maps extend into neighbourhoods of simplices, our definitions of the immersion and coplanar do not "see" tangent vectors that "point outside simplices."*

Let $f\colon |K| \to X$ be G-equivariant and X be a coordinate half-space. We require, for f to be regarded as a C^r map, that $f|\colon f^{-1}X' \to X' \hookrightarrow \mathbb{R}^n$ is coplanar.

Remark 46 *The property of f being a map (that is, $f|$ satisfying the coplanarity condition) is independent of the choice of X', since $D_x f(v) \in \partial X$ is equivalent to $D_x(g \circ f)(v) \in \partial X$ for any $g \in G$.*

Let $f\colon |K| \to M$, where M is a G-manifold with a boundary. For f to be regarded as a C^r map, we require that it is coplanar in the local coordinates.

Now we know what a C^r-map is. Next we define C^1-approximations. It turns out that even if all our maps are C^r, we want to study C^1-approximations, not C^r-approximations.

Definition 47 *Let f be a G-equivariant C^r-map $|K| \to X$, where X is a coordinate space but not a half-space and assume that $|K|$ is equipped with a simplexwise linear G-invariant metric. We say that a G-equivariant C^r-map $h\colon |K'| \to X$ is a δ-approximation to f, $\delta > 0$, if K' is an equivariant subdivision of K and the following conditions hold for all $x \in f^{-1}X'$:*

- $h(x) \in X'$.
- $|f(x) - h(x)| < \delta$
- $|D_x f(y-x) - D_x h(y-x)| < \delta |y-x|$ *for all* $y \in \mathrm{St}_x\, |K'|$.

Definition 48 *Let $f\colon |K| \to X$ be a G-equivariant C^r map, where X is a coordinate half-space. Let $K'' = f^{-1}X'$. Let $i\colon X' \hookrightarrow \mathbb{R}^n$. We say that a G-equivariant C^r-map $h\colon |K'| \to X$ is a δ-approximation to f if $i \circ h||K''|$ is a δ-approximation to $i \circ f||K''|$ and*

- *If* $i \circ f(x) \in \mathbb{R}^{n-1} \hookrightarrow H^n$ *then* $i \circ h(x) \in \mathbb{R}^{n-1} \hookrightarrow H^n$.

If

- $i \circ f(x) \in \mathbb{R}^{n-1}$ *if and only if* $i \circ h(x) \in \mathbb{R}^{n-1}$,

we say that h is a strong δ-approximation to f.

We say that $h\colon |K| \to \mathbb{R}^n$ is an overboard δ-approximation to f, if h satisfies all points in the definition of the boundary version of a δ-approximation to f, except that we do not require the condition $\operatorname{im} h \subset H^n$.

Remark 49 *Since $X = GX'$, our concept of approximation controls the behaviour of the entire h since it is equivariant, even though the definition formally speaks only about the behaviour in X'.*

Since X' has a fixed coordinate structure, the distance of two points is well-defined via that structure and the Euclidean norm. Furthermore, there is a canonical way to identify each $T_x X'$ with R^n, and thus, by using that canonical identification, the difference of two vectors in different tangent spaces is well-defined.

Also, the geometrical realization $|K| = |K'|$ of a simplicial complex K has a natural affine structure and a (simplexwise) linear metric, so that using $y - x$ as a tangent vector and speaking about its length are meaningful.

In the third condition, we actually need the Euclidean structure to move tangent vectors from one tangent space to another, and so just using an arbitrary equivariant Riemannian manifold on the target side is not enough, at least without heavy differential geometry.[1]

[1]The author wants to thank Pekka Pankka and Elena Ravaioli for helpful discussions concerning this remark.

In the non-equivariant case one can, by embedding the manifold into a Euclidean space, talk about approximations of functions from a (geometrical) simplicial complex into a manifold, and by embedding into a representation space also in the case of a finite G. However, in our case there is not necessarily a possibility to embed a manifold in an equivariant way into a Euclidean space, and thus we have to confine ourselves to approximating maps only inside chosen coordinate neighbourhoods. See Illman and Kankaanrinta [10] how this can be done in a wholesale way so that one obtains a topology in the function space (in the case of the paper of Illman and Kankaanrinta, into the space of maps between two differentiable manifolds).

Next, we plug the approximations inside coordinate neighbourhoods into the case of full manifolds.

Let (M, N) be a G-manifold pair, (K, K_0) be a locally finite G-complex pair, and $f\colon |K| \to M$ be a C^r-differentiable G-equivariant map. The map f is a map of pairs, $f\colon |K, K_0| \to (M, N)$, if f restricts to a map $f|\colon |K_0| \to N$.

Definition 50 *Let M be a G-manifold and $f\colon |K, K_0| \to (M, N)$ be a C^r, G-equivariant map. Let L be a G-invariant subcomplex of K, $X \subset M$ a coordinate neighbourhood and $\delta > 0$. Assume furthermore that $f \operatorname{St} |L| \subset X$, and that $L \cap f^{-1}X'$ is a finite complex.*

We say that $h\colon |K', K_0'| \to (M, N)$ is a (δ, L, X)-approximation to f, if for all $x \in |K|$ we have that

- *$f(x) \in X$ if and only if $h(x) \in X$*
- *$f(x) = h(x)$ if $x \notin L$, and the simplicial structure of K differs from the one of K' only inside L.*
- *$h| \operatorname{St} L$ is a δ-approximation to $f| \operatorname{St} L$ in X. (Distances measured in the Euclidean metric of X'.)*

If $h| \operatorname{St} L$ is a strong approximation to $f| \operatorname{St} L$, then we say that h is a strong approximation to f.

Acknowledgements

C^r-maps between simplicial complexes and C^r-manifolds are well-known. The coplanarity requirement is the author's addition, and it will prove useful when triangulating manifolds with a boundary without forming a double.

C^1-approximations are known, but Munkres [18] and Illman [7] use them in a context where one can embed the manifold into a coordinate space and thus they can define approximations globally through the norm of the coordinate space. Doing approximations locally and inside the special component of X' is the author's modification, inspired by Illman and Kankaanrinta [10].

The division between three different approximation concepts for manifolds with a boundary is added by me, and this is one aspect in which this thesis differs radically from my licentiate thesis [15]. In my licentiate thesis I simply formed the double of the manifold and triangulated it. Thus I did not have to care about boundaries at this level. Here I must take the boundary into the account on each level of construction.

Remark

Instead of C^1-approximations, it is possible to use Lipschitz approximations; these two approaches are equivalent for differentiable maps.

Assume K is a finite complex and $f\colon |K| \to \mathbb{R}^n$, and $f'\colon |K'| \to \mathbb{R}^n$, where K' is a subdivision of K. We say that f' is an ϵ-Lip-approximation to f, if $|f(x) - f'(x)| < \epsilon$ for all $x \in |K|$ and the map $f - f'$ is ϵ-Lipschitz.

The advantage of this approach is that one does not need to look into the simplicial structures of K and K', but it is enough to consider the space $|K| = |K'|$ with a simplexwise linear metric. I, however, chose to use the traditional C^1-approximations in this thesis, since this way I can refer to results in Munkres [18].

The use of Lip-approximations when triangulating manifolds was mentioned in Hirsch-Mazur [6], pp. 64-69, and was discovered independently by the author a couple of decades later.

2.2 Results on maps and approximations

We can do "copying with the action" also with simplicial complexes, as the following lemmas show. (Compare with Remark 23.)

Lemma 51 *Let $H < G$, and let (K_0, L_0) be an H-equivariant simplicial complex. Let $(K, L) = G \times_H (K_0, L_0)$, and let (K_0', L_0') be an H-equivariant subdivision of (K_0, L_0). Then $G \times_H (K_0', L_0')$ is a G-equivariant subdivision of (K, L).*

If, furthermore, the simplicial structures of K_0' and K_0 coincide inside some subcomplex K_0'' of K_0, then the simplicial structures of $G \times_H K_0'$ and $G \times_H K_0$ coincide inside $K'' = G \times_H K_0''$.

Proof: See Proposition 148 in Appendix A. □

Lemma 52 *Let (X, X_0) be a coordinate space pair. Let $f\colon |K, L| \to (X, X_0)$ be C^r, G-equivariant and $f'|$ be a $G_{X'}$-equivariant δ-approximation to $f|(f^{-1}X')$. Then $f'|$ extends* *to a G-equivariant δ-approximation to f. If $f|$ is a strong approximation to $f'|$, then f is a strong approximation to f'.*

Proof: The map f' is a δ-approximation to f if and only if $f'|f^{-1}X'$ is a δ-approximation to $f|f^{-1}X'$. Thus, in the case of a non-boundary coordinate space (X, X_0), it is enough to check that $f'|$ extends to an equivariant f'.

Let K_0 be the subcomplex of K whose geometrical realization is $f^{-1}X'$. Let K_0' be the domain of $f'|$, which is a subdivision of K_0. Then by Corollary 143 in the appendix, we have that $K = G \times_{G'_X} K_0$. Thus we get the invariant subdivision $K' = G \times_{G'_X} K_0'$ of K, and $f'|$ extends to an equivariant differentiable map $f' = G \times_{G'_X} f'|\colon |G \times_{G_{X'}} K'| \to X$ by the results in the appendix.

Assume then that X is a coordinate half-space. Let $x \in |K'| = |K|$, and let $g \in G$, $y \in f^{-1}X'$ be such that $gy = x$. Since G preserves ∂X, we have that $gf'(y) = f'(x) \in \partial X$ if and only if $f'(y) \in \partial X$, and $gf(y) = f(x) \in \partial X$ if and only if $f(y) \in \partial X$. Thus, if $f'|$ is an approximation to $f|$, then $f|(y) \in \partial X$ implies $f'|(y) \in \partial X$, which is equivalent to the claim that $f(x) \in \partial X$ implies $f'(x) \in \partial X$. Thus, if $f'|$ is an approximation to $f|$, then f' is an approximation to f.

If $f|$ is a strong approximation to $f'|$, a similar argument shows that f is a strong approximation to h.□

Lemma 53 *Let (X, X_0) be a coordinate neighbourhood in (M, N), let $f\colon |K, K_0| \to (M, N)$ be a G-equivariant C^r-map and L an invariant subcomplex of K such that $f|\operatorname{St} L| \subset X$. Let $f'|$ be a $G_{X'}$-equivariant $(\delta, L \cap f^{-1}X', X')$-approximation to $f|(f^{-1}X' \cap \operatorname{St} L)$. Then $f'|$ extends to a G-equivariant (δ, L, X)-approximation f' to f. If $f'|$ is a strong approximation to f, then also f' is a strong approximation to f.*

Proof: We have by Lemma 139 in the appendix that $\operatorname{St} L = G \times_{G_{X'}} (\operatorname{St} L \cap f^{-1}X')$. Thus, by Lemma 51, there exists a G-equivariant subdivision L' of $\operatorname{St} L$ such that $f'|\colon \operatorname{St} L \to X'$.

Since the simplicial structures of $L' \cap f^{-1}X'$ and $\operatorname{St} L \cap f^{-1}X'$ differ only inside L, that is, they coincide in $\partial \operatorname{St} L \subset \operatorname{St} L - L$, we have by Proposition 148 that the H-invariant subdivision $L' \cap f^{-1}X'$ extends to a G-invariant subdivision K' of K such that inside $\operatorname{St} L$ the structure of K' coincides with that of L' and outside $\operatorname{St} L$ with that of K.

Since $L' = G \times_{G_{X'}} (L' \cap f^{-1}X')$, we have that by Lemma 52 that $f'|\,|\operatorname{St} L \cap f^{-1}X'|$ extends to a (δ, L, X)-approximation $f'|\,|L'|$ to $f|\,|\operatorname{St} L|$, and that if $f'|\,|L' \cap f^{-1}X'|$ is a strong approximation to $f|$, then also $f'||L'|$ is a strong approximation to f.

We have that $f'|$ and $f|$ coincide outside L. Namely: $x \in |\operatorname{St} L| \setminus |L|$, then $x = gx'$, $x' \in |\operatorname{St} L \cap f^{-1}X'|$, and $(f'|\operatorname{St} L)(x) = f'|(gx') = gf'|(x') = gf(x') = f(gx')$.

Thus we can define $f'\colon |K', L'| \to (X, X_0)$ such that $f'(x) = f'|(x)$, if $x \in |\operatorname{St} L|$, and $f'(x) = f(x)$ otherwise. Since each $x \in |K'|$ has a neighbourhood U such that $U \subset |L'|$ or

$U \subset |K'| \setminus |L|$, this definition indeed gives us a C^r-map.□

We prove that our concept of approximation works well with respect to changing coordinates.

Proposition 54 [2] *Let K be a simplicial G-complex such that K/G is a finite complex. Assume that X_0 and X_1 are coordinate spaces, and U_0 an open invariant subset of X_0. Assume that $\phi\colon U_0 \to X_1$ and $f\colon K \to U_0$ are G-equivariant C^r maps.*

Then, for every $\epsilon > 0$, there exists $\delta > 0$ such that if h is a δ-approximation to f, then $\phi \circ h$ is an ϵ-approximation to $\phi \circ f$.

Proof: Let $\epsilon > 0$ be given. We can assume that $K \cap f^{-1}X_0'$ is connected. For, otherwise, we can repeat the proof for every subcomplex of the type GK_0, K_0 a connected component of $K \cap f^{-1}X_0'$, and take the minimum of the values of δ.

Let $K' = K \cap f^{-1}X_0'$. Since K' is connected, $(\phi \circ f)K' \subset X_1''$ for some connected component X_1'' of X_1. Let $g \in G$ be such that $gX_1'' = X_1'$ (the special component).

First, we choose δ so small that $hK' \subset \phi^{-1}X_1''$ for every h, h a δ-approximation to f.

We want to prove that shrinking δ further, we obtain that $g \circ \phi \circ h \circ g^{-1}|gK'$ and $g \circ \phi \circ f \circ g^{-1}|gK'$ satisfy the two inequalities in the definition of an ϵ-approximation. Since g^{-1} above is a simplexwise linear isometry, this is equivalent to the claim that $g \circ \phi \circ h|K'$ and $g \circ \phi \circ f|K'$ satisfy the inequalities in the definition of an ϵ-approximation.

Since $|K'|$ is compact, we find for $f|K'|$ a compact neighbourhood C in $U_0' = U_0 \cap X_0'$, and by diminishing δ, we can assume that $h|K'| \subset C$. Since $g \circ \phi$ is uniformly continuous on C, we can diminish δ so that $d(g \circ \phi \circ f(x), g \circ \phi \circ h(x)) < \epsilon$ whenever $d(f(x), h(x)) < \delta$. Thus the easier one of the inequalities in the definition of the δ-approximation is satisfied.

We prove that the other inequality is satisfied. Now,

$$|D_x(g \circ \phi \circ f)\bar{v} - D_x(g \circ \phi \circ h)\bar{v}| = |J_{g\circ\phi}(f(x))D_x f\bar{v} - J_{g\circ\phi}(h(x))D_x h\bar{v}| \leq$$
$$\|J_{g\circ\phi}(f(x))\|\,|D_x f\bar{v} - D_x h\bar{v}| + \|J_{g\circ\phi}(f(x)) - J_{g\circ\phi}(h(x))\|\,\|D_x h\|\,|\bar{v}|.$$

Here J denotes the Jacobian matrix, and the norm of the Jacobian is equivalent to the Euclidean norm when the space of matrices is considered as $\mathbb{R}^{n\times n}$. The symbol $\|D_x h\|$ denotes the fastest rate of growth of h at x. Here the domains of f and h are simplicial complexes, and thus, at the boundaries of simplices, we have to consider growth rates in different simplices, and take the maximum of these.

Now, $\|J_{g\circ\phi}(f(x))\|\,|C$ is a continuous function of $f(x)$, and since C is compact, $\|J_{g\circ\phi}(f(x))\||C$ is bounded by a constant, say M_0. We have that $\|D_x f\|\;|f^{-1}X'$ is upper semicontinuous as a function of x. Thus it is bounded by, say M_1', and since

[2]The author wants to thank Pekka Pankka for help with this proposition.

$|D_x f\bar{v} - D_x h\bar{v}| < \delta|\bar{v}|$, $\|D_x h\|$ is bounded by $M_1 = M_1' + \delta$. Now our formula above can be approximated from the above by

$$M_0|D_x f\bar{v} - D_x h\bar{v}| + M_1|J_{g\circ\phi}(f(x)) - J_{g\circ\phi}(h(x))||\bar{v}|,$$

and we want this formula to be less than $\epsilon|\bar{v}|$.

This can be obtained, as we diminish δ sufficiently:

- $\delta < \epsilon/(2M_0)$, this takes care of the first summand.
- Since $J_{g\circ\phi}$ is uniformly continuous on C, there exists δ' such that if $|f(x)-h(x)| < \delta'$, then $M_1|J_{g\circ\phi}(f(x)) - J_{g\circ\phi}(h(x))||\bar{v}|, < \epsilon/2$. Thus it is enough to shrink $\delta < \delta'$.

□

Next, we get to the main course of this section. We prove that if we have a map that is nice in certain ways, then it can be altered a little without ruining the niceness.

Proposition 55 *Let X be a coordinate space. Let $f\colon |K, L| \to (X, X_0)$ be a G-equivariant map such that $f^{-1}X'$ is a finite complex. Assume that f is an*

- *immersion*
- *embedding or*
- *Let A be an invariant subset of X such that $X' \cap A$ is compact. Let f be an embedding such that $A \subset\subset \operatorname{im} f$.*

Then there exists δ such that each δ-approximation h to f is

- *immersion*
- *embedding or*
- *embedding such that $A \subset\subset \operatorname{im} h$,*

respectively.

Proof: We restrict our attention to $f|\colon f^{-1}X' \to X'$. Then we have a finite complex, and the cases of immersion and embedding follow for $h|\colon f^{-1}X' \to X'$ from Munkres [18], Lemma 8.8. Then, the fact that the entire h is an immersion is proved as follows. Let $x \in |K|$, and let $g \in G$ such that $g|\operatorname{St} x| \subset f^{-1}X'$. Now $h|\,|\operatorname{St} x| = g^{-1}h|\,|\operatorname{St} gx|g$; thus $h|\,|\operatorname{St} x|$ is an immersion, as a composition of immersions, for all $x \in |K|$, and since the definition of an immersion is local, it follows that h is an immersion.

Also, injectivity inside the f-inverse image of X' follows as in Munkres [18], Lemma 8.8. Global injectivity follows by Lemma 16. Thus the claim about embeddings is true.

The last claim is seen as follows. Assume $A \subset\subset \operatorname{im} f$ is as in the claim. We claim that for small enough δ, $A \subset\subset \operatorname{im} h$. By enlargening A a little, we can assume that $A = \operatorname{cl}(A \setminus \partial X)$.

First we observe that it is possible to restrict our attention only inside the special component, and after the situation is handled inside it, the global claim follows by equivariance.

Our argument is essentially the same as in Munkres [18], Lemma 3.11, but in Munkres one has a surjective map, and he does not handle the boundary.

Let $(U_i \subset\subset V_i \subset\subset W_i)_i$ be three finite similar coverings of $A \cap X'$ with balls of X' that fit inside $\operatorname{im} f$. (Note that $\operatorname{im} f$ may meet the boundary of X in the case of a boundary neighbourhood, and in this case the balls are homeomorphic to Euclidean balls but they need not be isometric to Euclidean balls.)

For each ball B, we denote by ∂B the boundary of B in X'.

Let $\delta < \min_i(\min(d(\partial U_i, \partial V_i), d(\partial V_i, \partial W_i)))$.

Assume now that $x \in A \cap (X' \setminus \partial X')$ is not covered by $\operatorname{im} h$. Choose $U_i \ni x$. Now h determines a map $h' \colon (B^n, S^{n-1}) \to (W_i, (W_i \setminus U_i) \cup (W_i \cap \partial X'))$, where $(B^n, S^{n-1}) = f^{-1}(V_i, (\partial V_i) \cup (V_i \cap \partial X'))$. Now by pushing away from x, one easily constructs a map $h'' \colon (B^n, S^{n-1}) \to (W_i, (W_i \setminus U_i) \cup (W_i \cap \partial X'))$ such that $\operatorname{im} h'' \cap (U_i \setminus \partial X') = \emptyset$ and a homotopy of pairs $h' \sim h''$.

We can regard the map $f|(B^n, S^{n-1})$ as a map $f| \colon (B^n, S^{n-1}) \to (W_i, (W_i \setminus U_i) \cup (W_i \cap \partial X'))$. It is known from homotopy theory that $f|(B^n, S^{n-1}) \sim h'$ for a small enough δ (δ depending on i, not x), and, consequently, we have a homotopy of pairs $f|(B^n, S^{n-1}) \sim h''$. However, a simple homology argument proves that this is impossible. For, $H_n(W_i, (W_i \setminus U_i) \cup (W_i \cap \partial X'); \mathbb{Z}) = \mathbb{Z}$. We easily see that $H_n(f|)$ is surjective, but also $H_n(f|) = H_n(f'') = 0$, which is a contradiction.

Thus $A \cap (X' \setminus \partial X') \subset \operatorname{im} h$, and since $\operatorname{im} h$ is closed and $A \cap X' = \operatorname{cl}((A \setminus \partial X) \cap X')$, we have that $A \cap X' \subset \operatorname{im} h$. Since $A = G(A \cap X')$ and h is equivariant, $A \subset \operatorname{im} h$.

We had to shrink δ in the proof, but since the coverings $(U_i), (V_i), (W_i)$ are finite and the change of δ depends only on the chosen i and not on the point x, to run the argument through with all different x's we have to shrink δ altogether a finite number of times, once for each i. □

Actually, we are interested in using a modification of Proposition 55 for (δ, L, X)-approximations:

Proposition 56 *Let (X, X_0) be a coordinate neighbourhood and $f \colon |K, K_0| \to (M, N)$ be a G-equivariant map such that the inverse image of each component of X intersects only a finite number of simplices. Let $L \subset K$ be equivariant and such that $f|\operatorname{St}^2 L| \subset X$. Let furthermore f be an*

- *immersion*
- *embedding or*
- *embedding such that for an invariant closed* $A \subset M$ *we have that* $A \subset\subset \operatorname{im} f$.

Then there exists δ *such that each* (δ, L, X)*-approximation* h *to* f *is an*

- *immersion*
- *embedding or*
- *embedding such that* $A \subset\subset \operatorname{im} h$,

respectively.

Proof: Immersion: Being an immersion is a local property. Each point x of $|K|$ is an interior point of either $K \setminus L$ or $\operatorname{St} L$. Choose δ as in the previous proposition in the case of the immersion for the complex $\operatorname{St} L$. Then, in the first case, $f = h$ in some neighbourhood of x, and the claim is trivially true. In the case $x \in \operatorname{St} L$, the claim follows from Proposition 55 together with the choice of δ.

Imbedding: Similar to the previous case. Just choose δ so small that it is also smaller than the distance of fL and $X \setminus f \operatorname{St} L$, and also that the δ-neighbourhood of $f \operatorname{St} L$ is componentwise relatively compact and inside X. Then for no point $x \in L$ is it possible that $h(x) \in h(K \setminus \operatorname{St} L)$.

The last case: $(A \cap X) \setminus f|L|$ is always covered with $h|K \setminus L|$, and we simply have to make sure that $A \cap f|L|$ is covered with $h(|\operatorname{St} L|)$. Now $A \cap f|L|$ is componentwise compact (since L is a componentwise finite complex), and $A \cap f|L| \subset\subset f|K|$. Furthermore, X' intersects only a finite number of simplices, and thus there exists a neighbourhood U of $A \cap f|L|$ such that U intersects only simplices of $f|\operatorname{St} L|$. Thus, $A \cap f|L| \subset\subset f(|\operatorname{St} L|)$, and the required δ is given by Proposition 55. □

Acknowledgements

The main propositions, 55 and 56, are modifications of Munkres [18], Theorem 8.8 and Lemma 3.11. I modified the results in Munkres so that they fit in the equivariant and local-coordinate setting and to my special treatment of the boundary of a manifold.

2.3 Taking care of the boundary

In Illman [7], one triangulates manifolds with a boundary by triangulating the double of a manifold, and then gets the triangulation of the original manifold from the double. However, the author does not know how to form an equivariant double for a real analytic manifold. Therefore we triangulate manifolds with a boundary directly. This means that we have to take care of the boundary at each step of our constructions. Here we establish results that enable us to do that.

We aim at proving that near coplanar maps there are only coplanar maps, and that close enough overboard approximations to coplanar maps are actually strong approximations.

Lemma 57 *If $f\colon |K| \to \mathbb{R}^n$ is C^r and coplanar and s is a simplex in the barycentric subdivision of K, then there exists a face z of s such that for $x \in |s|$ we have $f(x) \in \mathbb{R}^{n-1}$ if and only if $x \in |z|$.*

Proof: Let s' be the simplex of K that contains s. Now s is of the form $\{v_1, \ldots, v_m\}$, where $v_1, \ldots, v_m$ are barycenters of faces $p_1, \ldots, p_m$ of s' such that p_i is a face of p_{i+1} for all $i \in \{1, \ldots, m-1\}$. Let i_0 be the greatest index such that $f(v_{i_0}) \in \mathbb{R}^{n-1}$. Since $f^{-1}\mathbb{R}^{n-1} \cap s'$ is a subcomplex of s', we have that $f(v_i) \in \mathbb{R}^{n-1}$ if and only if $i \leq i_0$. Since $f^{-1}\mathbb{R}^{n-1} \cap s'$ is a subcomplex of s', it follows that for $x \in |s|$ we have $f(x) \in f^{-1}\mathbb{R}^{n-1}$ if and only if $x \in |(v_1, \ldots, v_{i_0})|$.□

Lemma 58 *Assume s is a simplex, $f\colon |s| \to \mathbb{R}^n$ is C^r and coplanar and $f^{-1}\mathbb{R}^{n-1} = |z|$, z a face of s. Let b be the barycenter of z, and let v be a vertex of $s \setminus z$. If $x \in |z|$, then $D_x f(v-b) \notin \mathbb{R}^{n-1}$.*

Remark 59 *The vector $D_x f(v-b)$ is well-defined, since we can assume that f is extended into a map $U \to \mathbb{R}^n$, where U is a neighbourhood of $|s|$ in $\mathbb{R}^{\dim s}$.*

Proof of lemma:

Now $(b-x) = \sum t_i v_i$, where the v_i are vertices of z, and $t_i \in \mathbb{R}$. Thus, $D_x f(b-x) = \sum t_i D_x f(v_i) \in \mathbb{R}^{n-1}$.

If v is a vertex of $s \setminus z$, v does not lie on the hyperplane spanned by z. If $D_x f(v-b) \in \mathbb{R}^{n-1}$, then also $D_x f(v-x) = D_x f(v-b) + D_x f(b-x) \in \mathbb{R}^{n-1}$, violating the fact that f is coplanar.□

Lemma 60 *Assume $f\colon |s| \to \mathbb{R}^n$ is C^r, $f^{-1}\mathbb{R}^{m-1} = |z|$, where z is a face of s. Let b be the barycenter of z. If for all $x \in |z|$ and all vertices v of $s \setminus z$, we have that $D_x f(v-b) \notin \mathbb{R}^{n-1}$, then f is coplanar.*

Proof: Since $H^n \setminus \mathbb{R}^{n-1}$ is a connected component of $\mathbb{R}^n \setminus \mathbb{R}^{n-1}$ and $|s| \setminus |z|$ is connected, we can assume that $f\colon |s| \to H^n$.

Let v be a vertex of $s \setminus z$. Let U be a neighbourhood of $|s|$ such that f is defined in U.

Now $D_x f(v-x) = D_x f(v-b) + D_x f(b-x)$. Since $D_x f(b-x) \in \mathbb{R}^{m-1}$ and $D_x f(v-b) \notin \mathbb{R}^{m-1}$, we have that $D_x f(v-x) \notin \mathbb{R}^{m-1}$.

Now let $v' + x$ be an arbitrary point of $|s| \setminus |z|$. Now $v' + x = \sum t_i v_i$, v_i vertices of s, such that $t_{i_0} > 0$ for some i_0 such that $v_{i_0} \notin |z|$. Since $\sum t_i = 1$, we have that $v' = \sum t_i(v_i - x)$, and $D_x f(v') = \sum t_i D_x f(v_i - x)$. In the coordinates of H^n, the first (i.e., the one orthogonal to $\mathbb{R}^{n-1}$) coordinate of each $D_x f(v_i - x) \geq 0$, and the first coordinate of $D_x f(v_{i_0} - x) > 0$. Thus $D_x f(v') \notin \mathbb{R}^{m-1}$. □

Proposition 61 *Let $f\colon |s| \to \mathbb{R}^n$ be C^r, $f^{-1}\mathbb{R}^{m-1} = |z|$, z a face of s. Let $f'\colon U \to \mathbb{R}^n$ be an extension of f. Then f is coplanar if and only if for all $x \in |z|$ we have that $D_x f(v - b) \notin \mathbb{R}^{m-1}$ for all vertices v of $s \setminus s'$.*

Proof: Combine the two previous lemmas.□

In the definition of C^1-approximation one refers only to derivatives in directions that point inside the simplex one deals with. The next lemma lets us consider all derivatives.

Lemma 62 *Let $f\colon |s| \to \mathbb{R}^n$ be C^r and let $f'\colon U \to \mathbb{R}^n$ be a C^r-extension of f into some open subset. If $g\colon |s| \to \mathbb{R}^n$ is an ϵ-approximation to f, then, for some U', g extends to an ϵ-approximation $g'\colon U' \to \mathbb{R}^n$ to $f'|U'$.*

Proof: Since g is an ϵ-approximation to f and the simplex $|s|$ is compact, g is an ϵ'-approximation to f for some $\epsilon' < \epsilon$. Let g' be an extension of g to some neighbourhood. Let δ be so small that if $|x - y| < \delta$, then $|f'(x) - f'(y)| < (\epsilon - \epsilon')/2$, $|g'(x) - g'(y)| < (\epsilon - \epsilon')/2$, $|D_x f'v - D_y f'v| < (\epsilon - \epsilon')|v|/2$ and $|D_x g'v - D_y g'v| < (\epsilon - \epsilon')|v|/2$ for all tangent vectors v.

[The last two conditions can be satisfied due to the fact that for C^1-functions on compact sets partial derivatives are uniformly continuous, and so the difference of partial derivatives in x and y can be made arbitrarily small (in a uniform way) by choosing a small enough δ. Furthermore, it is easy to see that if the differences of all partial derivatives in x and y are less than ϵ'', then the quantity $|D_x f'v - D_y f'v|$ (resp., $|D_x g'v - D_y g'v|$) is less than $\epsilon''(\dim s)^2|v|$. The $(\dim s)^2$ coefficient is far from being optimal, but it nevertheless works.]

Let U' be the δ-neighbourhood of $|s|$. If $y \in U'$, then there exists $x \in |s|$, such that $d(x, y) < \delta$, and thus $|f'(y) - g'(y)| < |f'(y) - f(x)| + |f(x) - g(x)| + |g(x) - g'(y)| < \epsilon$.

If $y \in U'$, then there exists $x \in |\mathring{s}|$ such that $d(x, y) < \delta$. Since x lies in the interior of $|s|$, we have that $|D_x fv - D_x gv| < \epsilon|v|$ for tangent vectors in all directions, and thus $|D_y f'v - D_y g'v| \leq |D_y f'v - D_x fv| + |D_x fv - D_x gv| + |D_x gv - D_y g'v| < \epsilon|v|$. □

The rest of this section deals with proving that near coplanar maps there are only coplanar maps, and that close enough overboard approximations to coplanar maps are actually strong approximations.

We will follow—with appropriate modificatons—Munkres' [18] proof that near immersions there are only immersions. (Theorem 8.8 in Munkres [18].)

Definition 63 *Let s be a simplex, $x \in |s|$ and $f\colon |s| \to \mathbb{R}^n$ be C^r, $r \geq 1$. We say that $d_x f\colon |s| \to \mathbb{R}^n$ is the unique affine map such that*

$$f(y) = d_x f(y) + |y - x|\epsilon(y - x),$$

where $\epsilon(y - x) \to 0$, when $y \to x$.

Lemma 64 *Let $x \in |K|, f\colon |K| \to \mathbb{R}^n$. Then $d_x f\colon \operatorname{St} x \to \mathbb{R}^n$ is well-defined. If $\epsilon > 0$, there is a neighbourhood U of x, such that $d_x f|U$ is an ϵ-approximation to $f|U$.*

Proof: For each simplex s of $\operatorname{St} x$, such that $x \in |s|$, we have that $(d_x f)|\,|s|$ is well-defined. We have to check that the definitions in different simplices s,s' coincide. Thus, let s'' be the maximal common face of s, s'. Then $x \in |s''|$, and $(d_x f)|\,|s''|$ is the unique affine L such that $f(y) = L(y - x) + |y - x|\epsilon(y - x)$. Since L is unique, it must coincide with $(d_x f)|\,|s''|$(defined either via $|s|$ or $|s'|$.)

Let $\epsilon > 0$ be given. Since f and $d_x f$ are continuous and $f(x) = d_x f(x)$, there is a neighbourhood U of x such that if $y \in U$, then $|f(y) - d_x f(y)| < \epsilon$. Since $D_x f v = D_x d_x f v$ for all tangent vectors v and $d_x f$ and f have continuous partial derivatives in each simplex, there is a neighbourhood U of x such that $|D_y f v - D_y d_x f v| < |v|\epsilon$ for all $y \in U$ and all tangent vectors v.□

Lemma 65 *Let I be a simplicial complex consisting of one 1-simplex and two 0-simplices that triangulates the interval $[0, c]$, $c > 0$. Let $f\colon |I| \to \mathbb{R}^n$ be an affine map such that $f^{-1}\mathbb{R}^{n-1} = \{0\}$. There exists $\epsilon > 0$ such that the following holds:*

If I' is a subdivision of I, and $f'\colon |I'| \to \mathbb{R}^n$ is an ϵ-approximation to f such that $0 \in (f')^{-1}\mathbb{R}^{n-1}$, then $(f')^{-1}\mathbb{R}^{n-1} = 0$.

Let, furthermore, $\operatorname{pr}\colon \mathbb{R}^n \to \mathbb{R}$ be the orthogonal projection of $\mathbb{R}^n$ to the coordinate axis orthogonal to $\mathbb{R}^{n-1}$. Then we can choose $\epsilon = \operatorname{pr}(D_0 f(c)/(2c))$.

Proof: Let $\epsilon = \operatorname{pr}(D_0 f(c)/(2c))$. Since f is affine, we have that $\operatorname{pr}(D_0 f(c)/(2c)) = \operatorname{pr}(D_x f(c - x)/(2(c - x))$ for all $0 \leq x < c$.

Let now $f'\colon |I'| \to \mathbb{R}^n$ be an ϵ-approximation to f. Let $x \in I'$. Now

$$\operatorname{pr} \frac{D_y f'(c - y)}{c - y} > \operatorname{pr} \frac{D_y f(c - y)}{c - y} - \epsilon > 0$$

for all $0 \leq y \leq x$. Now, if $x > 0$, we have

$$d(f'(x), \mathbb{R}^{n-1}) = \int_0^x \operatorname{pr} \frac{D_y f'(c - y)}{c - y} dy > 0.$$

□

Lemma 66 *Let s be a simplex and a z face of s. Let b be the barycenter of $|z|$. Each $y \in |s|$ can be represented as $y = c_y + \sum t_i(v_i - b)$, where c_y is a point of $|z|$ and v_i are vertices of $s \setminus z$ and all $t_i \geq 0$.*

For points y of $|z|$, we have that $c_y = y$. Furthermore, the function $y \mapsto c_y$ is continuous.

Proof: Write $y = \sum t_i v_i + \sum t_j v_j$, where v_j are vertices of z, v_i are vertices of $s \setminus z$, $t_i, t_j \geq 0$ and $\sum t_i + \sum t_j = 1$. Now $y = \sum t_i(v_i - b) + \sum t_j(v_j - b) + b$. Thus we need to prove that $c_y = \sum t_j(v_j - b) + b \in |z|$. But $c_y = \sum t_j v_j + (1 - \sum t_j)b$. Now c_y is a weighted average of points in $|z|$, and $c_y \in |z|$, since $|z|$ is convex.

If $y \in |z|$, then each $t_i = 0$, and thus $y = \sum t_j(v_j - b) + b = c_y$.

The map $y \mapsto c_y$ is continuous, since $\sum t_j v_j + (1 - \sum t_j)b$ is continuous as a function of t_j's. □

Proposition 67 *Let $f\colon |s| \to H^n$ be C^r-differentiable and coplanar such that $f^{-1}\mathbb{R}^{n-1} = |z|$, where z is a face of s. Then there exists $\epsilon > 0$ such that*

1. *If $f'\colon |s'| \to \mathbb{R}^n$ is an overboard ϵ-approximation to f, then for all $x \in |z|$ and all $v \in |s| \setminus |z|$ we have that $D_x f'(v) \notin \mathbb{R}^{n-1}$.*
2. *There exists a neighbourhood U (depending only on f and ϵ) of $|z|$ in $|s|$ such that if f' is an ϵ-overboard approximation to f, then $U \cap f'^{-1}\mathbb{R}^{n-1} \subset |z|$.*
3. *If f' is an overboard ϵ-approximation to f, then f' is coplanar and a strong approximation to f.*

Proof: Let $\mathrm{pr}\colon \mathbb{R}^n \to \mathbb{R}$ be the orthogonal projection of $\mathbb{R}^n$ to the coordinate axis orthogonal to $\mathbb{R}^{n-1}$. Assume f has been extended into $f\colon U \to \mathbb{R}^n$.

We prove (1). By Lemma 61, $D_x f(v - b) \notin \mathbb{R}^{n-1}$ for all vertices v of $s \setminus z$. Since $\mathrm{pr}\,\frac{D_x f(v-b)}{|v-b|}$ is positive and a continuous function of x, there exists $\epsilon > 0$ such that $\mathrm{pr}\,\frac{D_x f(v-b)}{v-b} > 4\epsilon$ for all $x \in |z|$ and v a vertex of $s \setminus z$.

Now let f' be an ϵ-approximation to f, and make the antithesis that there exist $x \in |z|$ and a tangent vector v, $x + v \in |s| \setminus |z|$, such that $D_x f'(v) \in \mathbb{R}^{n-1}$. In particular, note that we can choose v as short as we like, and also that $d_x f'(x + v) \in \mathbb{R}^{n-1}$.

Choose a neighbourhood U of x such that $d_x f$ is an $\epsilon/2$-approximation to f in U and $d_x f'$ is an $\epsilon/2$-approximation to f' in U. The neighbourhood U exists by Lemma 64 (note that U depends on f'). Let $y \in |s|$. Represent $y = \sum t_i(v_i - b) + c_y$, where $c_y \in |z|$ and v_i are vertices of $s \setminus z$.

Let V be a convex neighbourhood of x in $|s|$ such that if $y \in V$, then $y, c_y \in U$. Without loss of generality, assume that v is so short that $x + v \in V$.

Let now L be a line segment joining $y = x+v$ and c_{x+v}. Consider $d_x f|L$. It is coplanar and satisfies $L \cap (d_x f^{-1})\mathbb{R}^{n-1} = c_{x+v}$. Furthermore, since $d_x f$ is affine, $D_{c_{x+v}} d_x f(v') = D_x d_x f(v') = D_x f(v')$, where v' is any tangent vector.

Let $x+v-c_{x+v} = \sum t_i v_i$, where v_i are vertices of $s \setminus z$. Now $\operatorname{pr} D_{c_{x+v}} d_x f(\sum t_i(v_i - b)) = \sum \operatorname{pr} D_{c_{x+v}} d_x f(t_i(v_i - b)) > 4\epsilon \sum t_i |v_i - b| \geq |x + v - c_{x+v}| 4\epsilon$. (The first equality follows, since $\operatorname{pr} D_x d_x$ is a linear function with all relevant values positive reals, and the last inequality is the triangle inequality. The middle inequality is the definition of ϵ together with the equation $D_x f(v_i) = D_{c_{x+v}} d_x(v_i)$.)

Now $d_x f'|L$ is an $\epsilon/2$-approximation to $f'|L$, which is an ϵ-approximation to $f|L$, which in turn is an $\epsilon/2$-approximation to $d_x f|L$. Thus, $d_x f'|L$ is an 2ϵ-approximation to $d_x f|L$. Since

$$\frac{\operatorname{pr} D_{c_{x+v}} d_x f(\sum t_i(v_i - b))}{2|x + v - c_{x+v}|} > 2\epsilon,$$

we have by Lemma 65 that $L \cap (d_x f')^{-1}\mathbb{R}^{n-1} = c_{x+v}$. Thus, $d_x f'(x + v) \notin \mathbb{R}^{n-1}$, which is a contradiction.

We prove (2). It is enough to prove the following: Given $x \in |z|$, there exist $\epsilon > 0$ and a neighbourhood V of x such that if f' is an ϵ-approximation to f, then $V \cap f^{-1}\mathbb{R}^{n-1} \subset |z|$. After that, we can take a finite cover of $|z|$ with such neighbourhoods, and the union of those neighbourhoods will be the required neighbourhood. ϵ will be the minimum of different ϵ's corresponding to the neighbourhoods.

Let $\epsilon > 0$ be as in (1), and let U be a neighbourhood of x such that $d_x f$ is an ϵ-approximation to f in U. Now $f'|U$ is a 2ϵ-approximation to $d_x f|U$. Let V be a convex neighbourhood of x such that if $y \in V$, then $c_y, y \in U$. Now similarly to (1) (but with f' instead of $d_x f'$), it follows that $f'^{-1}\mathbb{R}^{n-1} \cap V \subset |z|$.

We prove (3). Let ϵ be as in (1) and (2). Let U be the union of neighbourhoods obtained in (2). Now (1) and (2) prove that $f'|U$ is coplanar, and we get that f' is coplanar, once we choose $0 < \epsilon < d(\mathbb{R}^{n-1}, f(|s| \setminus U))$. Now $f'(|s| \setminus U) \subset H^n \setminus \mathbb{R}^{n-1}$. Since $H^n \setminus \mathbb{R}^{n-1}$ is a connected component of $\mathbb{R}^n \setminus \mathbb{R}^{n-1}$ and $|s| \setminus |z|$ is connected, we have that $f'(|s| \setminus |z|) \subset H^n \setminus \mathbb{R}^{n-1}$. □

Proposition 68 *Let $f\colon |K| \to \mathbb{R}^n$ be coplanar, where K is a finite complex. Then there exists $\epsilon > 0$ such that if f' is an overboard ϵ-approximation to f, then f' is a strong approximation and coplanar.*

Proof: Let K' be the barycentric subdivision of K. Now f can be regarded as a map $|K'| \to \mathbb{R}^n$.

By Lemma 57, for each simplex s of K', we have that $f||s|$ is such that $f^{-1}\mathbb{R}^{n-1}|s|$ is a face of s. Thus, by the previous proposition, we find an $\epsilon_s > 0$ such that any overboard ϵ-approximation to $f|\,|s|$ is coplanar and a strong approximation. Let ϵ be the minimum of such ϵ_s.

Let $f'\colon |L| \to \mathbb{R}^n$, where L is a subdivision of K and f' is an overboard ϵ-approximation to f. By Theorem 77, there exists a subdivision L' that is a common subdivision of L and K'. Now, if we regard $f' = f''\colon |L'| \to \mathbb{R}^n$, then f'' is an ϵ overboard approximation to f. By the choice of ϵ, f'' is a strong approximation to f and coplanar.

Now, f' is a strong approximation to f, since f' and f'' are the same as a map, and they differ only in the definition of simplices. Also, f' is coplanar, if $f'^{-1}\mathbb{R}^{n-1}$ is a subcomplex of L. But L is a subdivision of K, and $f'^{-1}\mathbb{R}^{n-1} = f^{-1}\mathbb{R}^{n-1}$ is a subcomplex of K, hence a subcomplex of L.□

Lemma 69 *Let $f\colon |K| \to X$ be a G-equivariant map, X a boundary neighbourhood. Let $x \in f^{-1}\partial X$, $y \in |\operatorname{St} x|$. If $D_x f(y - x) \notin T_{f(x)}\partial X$, then we also have $D_{gx} f(gy - gx) \notin T_{fg(x)}\partial X$ for all $g \in G$.*

Proof: If not, then $D_{gx} f(gy - gx) = \bar{v}$ for $\bar{v} \in T_{fg(x)}\partial X$. But now $D_x g^{-1} D_{gx} f(gy - gx) = D_x g^{-1}\bar{v} \in T_{f(x)}\partial X$, where $D_x g^{-1} D_{gx} f(gy - gx) = D_x f(y - x) \notin T_{f(x)}\partial X$, which is a contradiction.□

Proposition 70 *Assume $f\colon |K| \to X$ is a G-equivariant coplanar C^r-map.*

Then there exists $\epsilon > 0$ such that any G-equivariant ϵ-approximation $f'\colon |K| \to X$ to f is a strong ϵ-approximation and coplanar.

Proof: By Proposition 68, there exists $\epsilon > 0$ such that aly ϵ-approximation $f'|f^{-1}X'$ is coplanar and a strong approximation to f. Since $f'^{-1}\partial X = Gf'\partial X'$, we have that f' is a strong approximation. That f' is coplanar follows from the equation $f'^{-1}\partial X = Gf'^{-1}\partial X'$ and Lemma 69.□

Acknowledgements

This was inspired by the proof in Munkres [18] of the fact that near immersions there are only immersions.

Chapter 3

Finding good approximations

In the previous chapter, we introduced the concept of an ϵ-approximation and proved results about all ϵ-approximations. In this chapter, we will construct specific ϵ-approximations which will be useful in triangulating manifolds and proving the uniqueness result.

3.1 Rectilinear cell complexes

It turns out that good and easy-to-use approximations are simplexwise linear. Thus, we prove here results on simplexwise linear maps. There is nothing remarkably new in the results in this section, but we feel that giving exact proofs in our setting is worthwhile. We follow more or less Munkres [18], Section 7.

A cell complex is otherwise similar to a simplicial complex, but we allow arbitrary n-dimensional geometrical polyhedra, called cells, instead of n-dimensional simplices.

To be exact, a cell c is a compact subset of $\mathbb{R}^n$ determined by a finite set of inequalities

$$c = \{x \in \mathbb{R}^n \mid 0 \lessdot_i \ell_i(x), i = 0, \dots, m\},$$

where $\ell_i\colon \mathbb{R}^n \to \mathbb{R}$ are linear, and each $\lessdot_i$ is either $\leq$ or $=$. We notice that each cell is convex.

The dimension of a cell c is the dimension of the minimal hyperplane in $\mathbb{R}^n$ containing c. In $\dim c = m$, we call c an m-cell.

A face c' of the cell c is obtained by changing, none, some or all of the $\leq$-signs into $=$-signs in the definition. If a face c' is different from the original cell c, we say that c' is a proper face of c.A face consisting of only one point is called a vertex of the cell. We do not regard the empty set as a face.

Lemma 71 *We have the following:*

1. *Each face c' of c is contained in the boundary of c.*

2. *The intersection of two faces of c is either empty or a face of c.*

3. *Each cell c is the convex closure of its vertices.*

Proof: (1) See Definition 7.4 and Lemma 7.5 of Munres [18].
(2) Let c', c'' be two faces of c. Then $c''' = c' \cap c''$ is obtained from the set

$$\{x \in \mathbb{R}^n \mid 0 \lessdot_i \ell_i(x), i = 0, \ldots, m\}$$

defining c by changing each such $\lessdot_i$ into $=$ that is $=$ in the definition of either c' or c''. Thus c''' is a face of c', c''. Thus we have (2).

We prove (3). Induction: Obvious for 0-cells.

Let c be an n-cell. By Munkres [18], Lemma 7.3 and its proof, the boundary of c is the finite union of its $n-1$-faces. Now, as a convex compact set, c is the convex closure of its $n-1$-faces, and by the inductive assumption, each of the $n-1$-faces is a convex closure of its vertices. Thus c is the convex closure of the vertices of its $n-1$-faces, which are vertices of c. □

A cell complex K is a locally finite collection of cells c in $\mathbb{R}^n$ such that

- If c is a cell of K and c' is a face of c, then c' is a cell of K.

- For all $c_1, c_2 \in K$, we have that $c_1 \cap c_2$ is either empty or a face of c_1 and c_2.

An n-cell c is a simplex if it has $n+1$ vertices. In this case we have the abstract simplex s corresponding to c defined so that $s = \{v_0, \ldots, v_n\}$, where the $v_0, \ldots, v_n$ are the vertices of c.

Lemma 72 *In the above situation, the natural map $i\colon |s| \to c$ fixing each vertex and extending linearly is a bijection.*

Proof: By Lemma 71, (3), i is well-defined and surjective. The map i is injective, since the dimension of c is n, and thus the vertices v_i cannot lie in an $n-1$-dimensional hyperplane. □

A subdivision of a cell complex K is a cell complex K' such that $\bigcup_{c \in K} c = \bigcup_{c' \in K'} c'$, and each $c' \in K'$ is a subset of some $c \in K$. If each $c' \in K'$ is a simplex, then we say that K' is a simplicial subdivision of K. In this case there is an abstract simplicial complex K'' corresponding to K', so that both of them have the same vertices, and a natural isomorphism $|K''| \to |K'|$.

Later on, we will also need to take cellular subdivisions K' of simplicial complexes K. When we are doing this, we assume that the geometrical realization $|K|$ has been embedded simplexwise linarly into $\mathbb{R}^n$ so that it can be regarded as a cell complex (with $|s|$ being a cell for each simplex s of K), and thus the definition of a cellular subdivision above applies to it.

We study also equivariant rectilinear simplicial embeddings $f\colon |K| \to X$, where X is a coordinate space. By rectilinearity we mean that for each simplex s of $f|f^{-1}X'$, $f||s|$ is an affine map. Since f is completely determined by $f|f^{-1}X'$, it turns out that this "weak" type of rectilinearity is entirely adequate for our purposes.

Let K be a cell complex in $\mathbb{R}^n$, G a finite group and let $G \curvearrowright \mathbb{R}^n$ be linear. We say that K is G-complex, if G maps each cell of K onto a cell of K.

For non-equivariant rectilinear cell complexes, see Munkres [18] pp. 65-73, and for the equivariant case also Illman [7] pp. 200-206.

Lemma 73 (Equivariant variation of Munkres [18], Lemma 7.8) *Let K be a rectilinear G-cell complex in $\mathbb{R}^n$. Then there exists a G-equivariant rectilinear simplicial subdivision K' of K.*

Proof: The subdivision is constructed inductively with respect to n in n-dimensional skeletons K^n of K. Start with the 0-skeleton of K, and it is the triangulation of K^0. Assume that the $n-1$-skeleton is subdivided. Choose from the interior of each n-cell c a point x_c, and do that in an equivariant way, that is, x is a chosen point if and only if gx is a chosen point for each g.

This can be obtained as follows: Let c be a cell. Choose $x_c = \sum v_i/\#\{v_i\}$, where v_i are the vertices of s. This choice satisfies the condition in the previous paragraph, since the action takes a vertex into a vertex, and as a liner action, it commutes with taking the average.

Now we declare that the new simplices of K'^n are convex hulls of the previously constructed simplices on the boundary of c and the point x_c for each n-cell c of K. (Since each $g \in G_{X'}$ is linear, taking the convex hull is an equivariant operator. Note that this inductive step adds not only n-simplices but also new lower-order simplices. The geometrical realization of the new complex, however, coincides with $|K^n|$.) □

Lemma 74 (Modification of Illman [7], Lemma 1.2) *Let X be a coordinate space and $A \subset X$ a G-invariant componentwise compact subset. Then there exist a G-equivariant simplicial complex K such that K/G is finite and an equivariant rectilinear embedding $t\colon |K| \to X$ such that $A \subset\subset f|K|$.*

Proof: First take a (possibly non-equivariant) rectilinear triangulation of a compact set $B \subset X'$ such that $A \cap X' \subset\subset B \subset\subset X'$. This can be done, for example, by starting with the the fact that $X' \cong B(0, \epsilon)$ in $\mathbb{R}^n$, taking a cubic cell structure in $\mathbb{R}^n$ with small enough cubes (the diameter of the cube less than, say, $d(\partial A, \partial X')/3$), and letting L be the cell complex formed by cubes that fit inside X'.

Then, take the equivariant cell complex whose maximal-dimensional cells are all non-void cells of the type $\bigcap_{g\in G_{X'}} gs_g$, $s_g \in L$, and lower-order cells are their boundaries (in Munkres [18], proof of Lemma 7.7, it is stated that such an operation gives us a cell complex.).

Now, obviously, each of the intersections is a rectilinear convex cell, and since the interior of each $g\bigcup_{s\in L} s$ covers $A\cap X'$, the result is an $G_{X'}$-equivariant cell complex structure L' on a subset of X' whose interior covers $A \cap X'$. Then, use the preceding lemma on L' to obtain the required embedding $t'\colon |K'| \to X'$.

Now, extend it by the twisted product into $t = G\times_{G_{X'}} t'\colon |G\times_{G_{X'}} K'| \to X$. □

Remark 75 *If (X, X_0) is a coordinate space pair, the original cubic cell structure in the previous lemma can be chosen such that the intersection of $X_0 \cap X'$ with the cell system consists of entire cells (of the dimension of X_0), since $X_0 \cap X'$ is linear in the coordinates of X'. Since all the further constructions in the previous lemma only subdivide this cell structure, also the result is compatible with (X, X_0), that is, the resulting K is such that $f\colon (K, f^{-1}X_0) \to (X, X_0)$ is a map of pairs.*

Proposition 76 *Let K be a cell complex in $\mathbb{R}^n$ and G a finite group acting linearly on $\mathbb{R}^n$ such that $\bigcup_{c\in K} c$ is an invariant subset of $\mathbb{R}^n$. Then there exists a cellular G-equivariant subdivision K' of K.*

Proof: We immediately note that an intersection of two cells is a cell since cells are determined by a set of inequalities, and the set of inqualities determining the intersection is the union of the sets determining the original cells.

Now we let K' be the collection of cells of the type $\cap_{g\in G} gs_g$, s_g's cells of K. This collection is obviously G-invariant. In Munkres [18], proof of Lemma 7.7, it is stated that this kind of a collection is a cell complex.□

Theorem 77 *Let $f\colon |K| \to X$ and $h\colon |L| \to X$ be two G-equivariant rectilinear embeddings of simplicial complexes into a coordinate space such that their orbit spaces consist of a finite number of cells. Then they have equivariant subdivisions that intersect in a full subcomplex.*

Proof: Again, it is enough to prove the claim for $G_{X'}$-equivariant subcomplexes $K_0 = f^{-1}X'$ and $L_0 = h^{-1}X'$, and then use the twisted product construction to obtain the full case by Lemma 144.

By Munkres [18], Theorem 7.10, we have that K_0 and L_0 have subdivisions K_0' and

L'_0 such that

$$K'' = \{f|s| \mid s \text{ a simplex of } K'_0\} \cup \{h|s| \mid s \text{ a simplex of } L'_0\}$$

is a rectilinear (simplicial) cell complex. (Note that the language of Munkres omits the word rectilinear, but it is implicitly present. See the proof of the cited Theorem 7.10.) Use the previous proposition to subdivide K'' further into a G-equivariant cell subdivision and then Lemma 73 to get an equivariant simplicial subdivision of K''. Now $f^{-1}K''$ and $h^{-1}K''$ are the required subdivisions that intersect in a subcomplex. Take barycentric subdivisions to guarantee intersection in a full subcomplex.□

Corollary 78 *The previous theorem can be strengthened so that K' and L' coincide with K and L, respectively, outside the second star of the intersecting part.*

Proof: Assume again that $K = f^{-1}X'$, $L = f^{-1}X'$. Once we have dealt with this case, extend to the full case by the twisted product construction.

Let $K_0 = \operatorname{St} f^{-1}(f|K| \cap h|L|)$, $L_0 = \operatorname{St} h^{-1}(f|K| \cap h|L|)$. Then use the previous proposition to obtain K'_0, L'_0 that intersect in a full subcomplex. Use the standard extension, Definition 7.12 in Munkres [18], to extend these subdivisions into subdivisions K_0 and L_0 that intersect in a subcomplex.

We have that K'_0, L'_0, K and L are equivariant complexes, and the standard extension procedure described in Munkres [18] refers to affine operations, simplices of K'_0, L'_0, K, L, plus taking the barycenter of some simplex. Thus, the linearity of the action guarantees that the standard extension is an equivariant subdivision. □

Remark 79 *Since we just subdivide and subdivision does not destroy the pair structure, the previous theorem and corollary are true also for maps of pairs $(K, K_0) \to (X, X_0)$, $(L, L_0) \to (X, X_0)$ (meaning that K'_0 and L, as well as L'_0 and K intersect in a subcomplex.)*

Lemma 80 *Let $f\colon |K| \to X$ be a G-equivariant rectilinear injective map, K a simplicial G-complex. Then f is a (coplanar) embedding.*

Proof: If X is not a half-space, all we have to prove is that f is an immersion at each point of $|K|$. Since a linear map is an immersion, all we are left to check is that the angle of two simplices is positive.

Make the antithesis that the there exist $x \in |s|, |s'|$, s, s' simplices of K, and rays r, r' starting at x and contained in $|s|, |s'|$, respectively, such that $r \cap |s'| = \{x\}$ and $r' \cap |s| = \{x\}$, but $D_x fr$ and $D_x fr'$ intersect at $y \in T_{f(x)}M, y \neq 0$. By equivariance, we may assume that $x \in f^{-1}X'$.

Since $D_x f$,r and r' are linear, we can choose y arbitrarily small. For small enough y, there exist $z \in s \cap r$ and $z' \in s' \cap r'$ such that $D_x f(z - x) = D_x f(z' - x) = y$. But now, since f is linear in s and s', we have that $f(z) = f(z') = x + y$, a contradiction with the injectiveness of f.

Assume then that X is a coordinate half-space. Since all geometrical simplices $f|s|$ are rectilinear, $f^{-1}\partial X$ is a subcomplex of K. The derivative condition is also easily seen to hold: Assume, namely, that $D_x f(v) \in \partial X', x \in \partial X'$. Then $f(v) = D_x f(v) + f(x) \in \partial X'$. □

Now we have proved that there exist rectilinear simplicial complexes triangulating coordinate neighbourhoods, and that rectilinear simplicial complexes intersect in a subcomplex after subdivision. The rest of the chapter deals with proving that C^r-maps from simplicial complexes to coordinate spaces have arbitrarily accurate approximations which are simplexwise linear.

Let $|K|$ be an equivariant simplicial complex with an invariant linear metric. Following Munkres [18], we define that the thickness of a simplex s is

$$\frac{\min_{v \text{ is a vertex of } s} d(v, b(s))}{\operatorname{diam} s}.$$

Here $b(s)$ is the barycenter of s.

Lemma 81 *Assume that K is a simplical G-complex such that $|K/G| = |K|/G$. Then $\pi\colon K \to K/G$ is injective in each simplex. Furthermore, a subdivision K'' of K/G induces an equivariant subdivision K' of K such that $\pi'\colon K' \to K''$ is simplicial.*

Proof: Assume $|K/G| = |K|/G$, and let s be an n-simplex in K. Since the action of G is properly discontinuous and linear in each simplex, one easily sees that $\pi|s| \subset |K|/G$ is n-dimensional. Namely, if $\pi|s|$ was less than n-dimensional, then an inverse image of a point in $\pi|s|$ should have a positive dimension, and it would be in an orbit with cluster points, which is impossible with a properly discontinuous action.

But $\pi|s| \subset |K/G|$ is n-dimensional if and only if $\pi|s$ is injective. Thus $\pi|s$ is injective for each simplex s of K.

Let $t\colon |K''| \to |K/G|$ be a subdivision map. Let K' be the complex whose vertices are the points $x \in |K|$ such that $t^{-1}\pi x$ is a vertex of $|K''|$, and $s = \{v_0, \ldots, v_n\}$ is a simplex of K' if and only if $t^{-1}\pi s$ is a simplex of K'' and all points $v_0, \ldots, v_n$ are contained in a simplex of $|K|$. Since the map π is invariant, we easily see that the set of vertices of K' is G-invariant subset of K, and the induced action of G in K' is simplicial.

Since the vertices of K' are points of $|K|$ and each simplex of K' is contained in a simplex of $|K|$, the inclusion $K' \subset |K|$ extends to a simplexwise linear G-equivariant map $t'\colon |K'| \to |K|$. We claim that t' is a subdivision of $|K|$. We thus need to check that it is bijective.

Let $x \in |K|$. Let $|s|$ be a simplex of $|K|$ containing x. Then $\pi|s|$ is a simplex $|K/G|$, and so there exists a simplex s' of K'' such that $\pi x \in t|s'|$ and $t|s'| \subset |s|$. Let $v_1, \ldots, v_n$

be the vertices of s', and $\pi x = t \sum \ell_i v_i$. Now let for each i, v'_i be the point of $|s|$ such that $\pi v'_i = t v_i$. We have, by the definition of K', that $\{v'_0, \dots, v'_n\}$ is a simplex of K', and we easily see that $x = t'(\sum \ell_i v'_i)$. Thus t' is surjective.

To prove injectivity, it is enough to check that $t'|t'^{-1}|s|$ is injective for each simplex s of K. To prove this, it is enough to prove that $(t^{-1} \circ \pi \circ t')|t'^{-1}|s|$ is injective. But one easily sees that $(t^{-1} \circ \pi \circ t')|t'^{-1}|s|$ is a simplicial isomorphism of abstract complexes and thus injective on the level of realizations. Since $\pi' = (t^{-1} \circ \pi \circ t')$, the previous observation also proves that π' is simplicial. □

Proposition 82 (Similar to discussion in Illman [7] p. 211) *Let K be a simplicial G-complex with an invariant metric linear in each simplex and a finite orbit complex. Then there exists $t > 0$ such that for any $\epsilon > 0$ there exist equivariant subdivisions of K having minimum thickness of simplices $> t$ and mesh $< \epsilon$.*

Proof: Start by replacing K with its second barycentric subdivision. This ensures that $|K|/G$ and $|K/G|$ coincide.

By Munkres [18], the beginning of the proof of the Lemma 9.4, the truth of our lemma is independent of the choice of the invariant metric in $|K|$. Thus we can assume that $|K|$ and $|K/G|$ have the standard metric defined in the Section 0.3 of this thesis. Now we easily see that $\pi\colon |K| \to |K/G|$ preserves distances inside each simplex s of $|K|$.

We apply the non-equivariant version of the lemma (for proof, see Munkres [18], Lemma 9.4) on the orbit space level, and this induces the required subdivision for K by the previous lemma.□

If K is a simplical complex, K' a subdivision of K and $f\colon |K| \to X$ equivariant, then $L_f\colon |K'| \to X$ is the equivariant map for which $L_f(v) = f(v)$ for all vertices v of K', and $L_f|\,|s|\colon |s| \to X'$ is linear in each simplex s of K' such that $f|s| \subset X'$. Since X' is a convex set, L_f is well-defined in X', and by the twisted product construction, $L_f|f^{-1}X'$ extends uniquely into $L_f\colon |K'| \to X$.

We denote by $\operatorname{mesh} K'$ the supremum of diameters of simplices in K'.

Theorem 83 *Let $f\colon |K| \to X$ be a G-equivariant C^r-embedding such that each component of X contains only a finite number of simplices. Let δ and t be given positive numbers. Then there exists $\epsilon > 0$ such that if for a G-equivariant subdivision K' of K we have that* $\operatorname{mesh} K' < \epsilon$ *and the minimum simplex thickness of K' is $> t$, then $L_f\colon K' \to X$ is a δ-approximation to f.*

Proof: If X is not a half-space, it is enough to find ϵ_s for every simplex s of $f^{-1}X'$ such that $L_f|s$ is a δ-approximation $f|s$. Then we can choose $\epsilon = \min \epsilon_s$.

But the existence of ϵ_s follows from Munkres [18], Lemma 9.3.

Assume X is a half-space. We have to check that $f(x) \in \partial X' \hookrightarrow H^n$ implies $L_f(x) \in \partial X' \hookrightarrow H^n$. Since f is an embedding, $f^{-1}\partial X$ consists of entire simplices. Assume $f(x) \in \partial X'$, and let s be a simplex such that $x \in |s|$, and $f|s| \subset \partial X'$. Now $L_f(x)$ is a weighted average of points $f(v)$, v a vertex of s. Thus the points $L_f(v) = f(v) \in \partial X'$, and since $\partial X'$ is affine, $L_f(x) \in \partial X'$.

By Lemma 80, L_f is coplanar. □

By combining the previous proposition and theorem, we get

Corollary 84 *Given a G-equivariant embedding $f\colon |K| \to X$ and $\delta > 0$, there exists a G-equivariant $h\colon |K'| \to X$ such that h is a δ-approximation to f and it is linear in each simplex.*□

The case of pairs requires virtually no extra work.

Corollary 85 *Given a G-equivariant C^r-map $f\colon |K, L| \to (X, X_0)$ as before, and $\delta > 0$, there exists a G-equivariant $h\colon |K', L'| \to (X, X_0)$ such that h is a δ-approximation to f and it is linear in each simplex.*□

Proof: The h given by the previous corollary is good enough for us, since both h and X_0 are linear, and $h(v) \in X_0$ for each vertex v of L'.□

Acknowledgements

Here we just restate results from Munkres [18], with minor modifications to deal with equivariance and coplanarity. Lemma 81 was used without proof by Illman [7]; the proof is my own.

3.2 The Extension Lemma

We are given an equivariant C^r map $f\colon |s| \to X$ and a really accurate approximation $f'|$ to $f||\dot{s}|$. We want to extend f' into an accurate approximation $f'\colon |s| \to X$, which is C^r.

It turns out that this is one of the most difficult parts of the triangulation results. This is exactly the place where Munkres made modifications to Whitehead's original lemma so that it no longer worked in the real-analytic case. Furthermore, here also the step from the non-equivariant case to the equivariant case requires a non-trivial argument, which we borrow from Illman [7], Lemma 2.5.

Definition 86 *Let $f\colon |s| \to \mathbb{R}^n$, and $f'\colon |s'| \to \mathbb{R}^n$ be an approximation to f. We say that f' preserves affine spaces, if for each affine $A \subset \mathbb{R}^n$ it is true that if $f|s|, f'|\dot{s}| \subset A$, then also $f'|s| \subset A$.*

Lemma 87 (Extension Lemma) Modification of Whitehead [24], Theorem 2 and Shiota [22], Lemma 6.8

Let s be a simplex and $t\colon |s| \to \mathbb{R}^n$ a C^r-map. Let $\epsilon > 0$. Then there exists $\delta > 0$ such that any δ-approximation $t''\colon |\overset{\bullet}{s}'| \to \mathbb{R}^n$ to $t||\overset{\bullet}{s}|$ can be extended to a C^r ϵ-approximation $t'\colon |s'| \to \mathbb{R}^n$ to t.

Furthermore, t' can be chosen so that it preserves affine spaces.

Proof: Let the dimension of s be m. We embed $|s|$ into $\mathbb{R}^m$ via a linear distance-preserving embedding $e\colon |s| \to \mathbb{R}^m$ so that the barycenter of $|s|$ goes to 0. We identify $|s|$ with the cell complex in $\mathbb{R}^m$ whose cells are the sets of the type $e|f|$, f a face of s. Let $u\colon |\overset{\bullet}{s}'| \to |\overset{\bullet}{s}|$ be the subdivision map. We identify $|\overset{\bullet}{s}'|$ with the cell complex in $\mathbb{R}^m$ consisting of images of simplices of $\overset{\bullet}{s}'$ in the mapping $e \circ u$.

We describe first the subdivision of s that yields s'.

First, we give a cell subdivision s'' of s. For each $m-1$-simplex z of $\overset{\bullet}{s}'$, we declare that the sets

$$\{x \in \mathbb{R}^m | x = \ell y, y \in |z|, 0 \le \ell \le 1/2\} \tag{3.1}$$

and

$$\{x \in \mathbb{R}^m | x = \ell y, y \in |z|, 1/2 \le \ell \le 1\} \tag{3.2}$$

are m-cells. We call 3.1 and 3.2 the cells of the first and the second type, respectively.

It is easy to see that these are indeed m-cells, and that they subdivide s. Furthermore, at the boundary the cell structure coincides with the one of $\overset{\bullet}{s}'$. By a known result (see Rourke-Sanderson [21], Proposition 2,9), we can further subdivide these cells into simplices in such a manner that the subdivision process does not subdivide any of the simplices at the boundary of s. Call the result complex s'. Since the further subdivision only relaxes the property of a map being C^r, it is enough to define a C^r-map $t'\colon |s''| \to \mathbb{R}^n$, and this gives us the desired map C^r-map $t'\colon |s'| \to \mathbb{R}^n$.

We define t' as follows: For each cell z of the first type, we let $t'||z| = t||z|$. For each cell z of the second type, we let z' be the $m-1$-cell at $z \cap \overset{\bullet}{s}'$. For each $y \in z'$ and each $1/2 \le \ell \le 1$, we let $t'(\ell y) = t(\ell y) + 2(\ell - 1/2)(t'(y) - t(y))$. We readily see that our different definitions in different cells coincide in the commont boundaries of cells, and t' is indeed a cellular map. We, furthermore, see that the definition of t' extends the definition of t'' at $\overset{\bullet}{s}'$.

We prove that t' is C^r. Inside cells of the first type this is obvious, since t' coincides with the C^r-map t. Let then z be a cell of the second type and z' be the corresponding $m-1$ cell at $\overset{\bullet}{s}'$. Let z'' be a neighbourhood of z' in the minimal-dimensional hyperplane that contains z' such that the map t can be extended to a C^r-map $t\colon z'' \to \mathbb{R}^n$. Let c be a small positive constant.[1]

Let $f\colon |z''|\times]1/2-c, 1+c[\to \mathbb{R}^m$ be the map $f(y,\ell)\mapsto \ell y$. For $c<1/2$ we see that f is a real-analytic embedding of $z''\times]1/2-c,1+c[$ onto a neighbourhood of $|z|$.

Furthermore, let $t'''\colon |z''|\times]1/2-c,1+c[\to \mathbb{R}^n$, $t'''(y,\ell)=t(\ell y)+2(\ell-1/2)(t'(y)-t(y))$. Now $t'|z=t'''\circ f^{-1}|z$, and to prove that t' is C^r, it is enough to prove that t''' is C^r. But this is obvious, since t''' is a composition of the C^r-maps $t, t', +, \mathrm{pr}_i$.

We must still prove the δ-ϵ -thing. For cells of the first type this is again obvious, since there $t=t'$. Let $\epsilon>0$. If we choose $\delta<\epsilon$, we have for cells of the second type that $d(t'(\ell y),t(\ell y))<2(\ell-1/2)d(t'(y),t(y))<d(t'(y),t(y))<\delta$.

Thus we are left with the derivatives. For $x\in|z|$, write $x=\ell y$, where $y\in|\overset{\bullet}{s}|$, and define C^r functions $\ell(x)=\ell$ and $y(x)=y$. Now we can write

$$t'(x)=t(x)+2(\ell(x)-1/2)(t'(y(x))-t(y(x)),$$

and $|D_x(t'(x)-t(x))|=|D_x(2(\ell(x)-1/2)(t'(y(x))-t(y(x)))|=$

$$|2\frac{\partial\ell(x)}{\partial x}(t'(y(x))-t(y(x)))+2(\ell(x)-1/2)\frac{\partial(t'(y)-t(y))}{\partial y}\frac{\partial y}{\partial x}|. \qquad (3.3)$$

Now, we easily see that $|\frac{\partial\ell(x)}{\partial x}|\le c'$, where c' is the inverse of the distance of $|\overset{\bullet}{s}|$ from the origin. Moreover, $|\frac{\partial y}{\partial x}|\le 2$, since the coefficient $1/2\le\ell\le1$. Thus we can approximate 3.3 from above by

$$2c'|t'(y(x))-t(y(x))|+2|\frac{\partial(t'(y)-t(y))}{\partial y}|. \qquad (3.4)$$

Assume then $t'(y)$ is a δ-approximation to $t(y)$, where $\delta<\epsilon/4, \delta<\epsilon/(4c')$. Then the formula 3.3 is at most ϵ.

We prove that t' preserves affine spaces. Let $A=\{x|\ell(x)\ge0\}$, and let $t|s|, t'|\partial s|\subset A$. Then $t'(x)$, $x\in|s|$ is a of the form $t(y)-c''(t'(y')-t(y'))$, where $t(y), t'(y'), t(y')$ all lie in A. Thus $t'(x)\in A$. □

The above-stated Extension Lemma works only for maps $|s|\to\mathbb{R}^n$. We do not have any a priori guarantee that extensions of maps $|s|\to H^n$ would stay inside H^n. However, preservation of affine spaces allows us to deduce that for accurate enough approximations the extension remains inside H^n. We show this with the following two corollaries.

Corollary 88 *If $t\colon|s|\to H^n$ is coplanar and $t'|\overset{\bullet}{s}$ is a strong δ-approximation to $t|$, then, in the previous lemma, t' is an overboard ϵ-approximation to t.*

Proof: Case (1): $t|s|\subset\mathbb{R}^{n-1}$. Since $t'|$ is a strong approximation to $t|$, we have that $\operatorname{im} t'|\subset\mathbb{R}^{n-1}$. But since t' preserves affine spaces, $t'|s|\subset\mathbb{R}^{n-1}$.

[1] We need this trickery of getting a neighbourhood of the domain because of the definition of the map being C^r.

Case (2): $t|s| \not\subset \mathbb{R}^{n-1}$. Since $t^{-1}\mathbb{R}^{n-1}$ is a subcomplex of $|s|$, $t\mathring{s} \cap \mathbb{R}^{n-1} = \emptyset$. Thus t' is an overboard ϵ-approximation to t if and only if $t'|$ is an overboard approximation to $t|$, but this is true by assumption.□

Corollary 89 *Let s be a simplex and $t\colon |s| \to H^n$ a coplanar C^r map. Let $\epsilon > 0$. Then there exists $\delta > 0$ such that any strong δ-approximation $t''\colon |\overset{\bullet}{s'}| \to H^n$ to $t||\overset{\bullet}{s}|$ can be extended to a strong ϵ-approximation $t'\colon |s'| \to H^n$ to t.*

Proof: By Proposition 68, we can shrink ϵ so that each overboard ϵ-approximation t' to t is a strong approximation. Use Lemma 87 to find δ for this ϵ. By the previous corollary, t' is an overboard approximation to t, and by the choice of ϵ, t' is a strong approximation to t. □

Next we show how one transfers from the non-equivariant case to the equivariant case.

Lemma 90 (Modification of Illman [7], Lemma 2.5) *Let s be an equivariant regular simplex, and $f\colon |s| \to X$ a G-equivariant-C^r map, X a neighbourhood that is not a boundary neighbourhood. Let $\epsilon > 0$. There exists δ such that if $f''\colon |\overset{\bullet}{s'}| \to X$ is a G-equivariant C^r δ-approximation to $f||\overset{\bullet}{s}|$, then f'' can be extended into a G-equivariant C^r $f'\colon |s'| \to X$ that is an ϵ-approximation to f.*

Furthermore, f' can be chosen so that it preserves affine spaces.

The above is true also in the situation where X is a boundary neighbourhood, f is coplanar, and we require f' to be a strong ϵ-approximation and coplanar.

Proof: Assume s is an n-simplex. Let z be an n-simplex in s such that $f|z| \subset X'$. Let $H_z < G$ be the subgroup that fixes z. (Since s is a regular simplex, either $g \in G$ fixes s pointwise or $g\mathring{s} \cap \mathring{s} = \emptyset$.)

Let $Y \subset X'$ be the subspace that is fixed pointwise by H_z. Since H_z fixes z pointwise and f is equivariant, $f|z| \subset Y$. For the same reason, $f'|\overset{\bullet}{z}| \subset Y$. Now $X' \cong B(0,t)$ (Euclidean ball, or a ball in H^n), and $Y = L \cap B(0,t)$, where L is a linear subspace in $\mathbb{R}^n$. Thus, given $\epsilon > 0$, we can find $\delta > 0$ such that if $f'||\overset{\bullet}{z'}|$ is a δ-approximation to $f|$, then it can be extended into $f'||z'|$ that is an ϵ-approximation to $f|$ by Lemma 87, or in the case of a boundary neighbourhood, by corollaries following Lemma 87. Since f' preserves affine subspaces, we have $f'||z'| \subset Y$, which means that $f'||z'|$ is an H-equivariant map (since both the domain and the image have the trivial action of H.)

We see that $|\mathring{s}|$ and $|\mathring{z'}|$ satisfy the assumptions of Lemma 11, and thus we have $|\mathring{s}| = G \times_{H_z} |\mathring{z}|$. We define $f'||\mathring{s}| = G \times_{H_z} f'||\mathring{z}|$, and $f'|||\overset{\bullet}{s}| = f''$. From Lemma 149, applied to z' and s, we get a subdivision s' of s that is the domain of f'.

We claim that f' is the desired ϵ-approximation. It is obviously G-equivariant. We prove that f' is C^r. Let z' be an n-simplex of z, and gz' an arbitrary n-simplex of s'.

Then $f'|gz'| = gf'|z'|$, which is C^r. Assume now that $fgz' \subset X'$. Let $gx \in |gz'|$ be an arbitrary point of $|s'|$. Then $d(f(gx), f'(gx)) = d(gf(x), gf'(x)) = d(f(x), f'(x)) < \epsilon$.

Now let $x \in |s|$ such that $x \in X'$. Let $y \in \mathrm{St}_{s'} x$. Since $s = Gz$, and x, y lie in a same simplex of $|s|$, there exists $g \in G$ such that $x' = gx$ and $y' = gy$ both lie in $|z|$. Obviously, $g \in G_{X'}$.

Now $f(x) = gfg^{-1}(x')$ and $f'(x) = gf'g^{-1}(x')$, and similarly for y. By Jacobian matrixes, we get $D_x f(y - x) = D_{f(x')} g D_{x'} f D_x g^{-1}(y - x) = D_{f(x')} g D_{x'} f(y' - x')$, and similarly for f'. Now

$$|D_x f(y - x) - D_x f'(y - x)| = |D_{f(x')} g D_{x'} f(y' - x') - D_{f'(x')} g D_{x'} f'(y' - x')| =$$

since g is an orthonormal map

$$|gD_{x'} f(y' - x') - gD_{x'} f'(y' - x')| = |D_{x'} f(y' - x') - D_{x'} f'(y' - x)| < \epsilon|y' - x'| = \epsilon|y - x|.$$

We prove that f' preserves affine spaces. Let A be an affine space such that $f(gz) \subset A$ and $f''(g\partial z) \subset A$. Now we have $f(z), f''(\partial z) \subset g^{-1}A$, where $g^{-1}A$ is also an affine space, since the action is orthogonal.

We can assume that $f'||z|$ was chosen so that it preserves affine spaces, and thus $f'|z| \subset g^{-1}A$. Thus $f'|gz| \subset A$.

Assume f is coplanar. Then, by shrinking ϵ, we obtain the result that f' is coplanar and a strong approximation by Proposition 70. □

Corollary 91 *Lemma 90 is true also for maps of pairs.*

Proof: Let $\epsilon > 0$ be given. We claim that the same δ as in Lemma 90 works also in this case. In this situation, f is a map $|s', s_0'| \to (X, X_0)$, where X_0 is a linear subspace and s_0' is an invariant subcomplex of s', and $f'|$ is a δ-approximation to $f|$. By Lemma 90, we get an ϵ-approximation $f'\colon |s'| \to X$. Since f' preserves affine spaces and $f(f^{-1}X' \cap s_0'), f'|(f^{-1}X' \cap \partial s_0') \subset X_0'$, we have that $f'(s_0' \cap f^{-1}X') \subset X_0'$. Thus, by the equivariance of f', we have that $f's_0 \subset X_0$. That is, f' is a map of pairs. □

Acknowledgements

Lemma 87 is due to Whitehead [24], but the observation that it preserves affine spaces is my own. Lemma 90 is due to Illman [7] for a finite G and C^∞-action. Extending it to cover the case of a properly discontinuous action is my own work, as well as combining it with Lemma 87 to cover the C^ω-case. The use of preservance of affine spaces to cover the case of boundary neighbourhoods is my own addition. The pair version, Corollary 91, was proved by Illman [7] with a slightly different technique.

3.3 Extending approximations and fitting complexes together

At the end of Section 3.1 we proved that every $f\colon |K| \to X$, where $f^{-1}X'$ is a finite complex, can be approximated arbitrarily finely by a piecewise linear map. In this section we will improve the result as follows.

We are given an equivariant map $f\colon |K| \to M$, where M is a manifold. We are, furthermore, given $x \in M$ and a coordinate neighbourhood X for x. We are still given a subcomplex K_0 of K such that $f(|\operatorname{St}^2 K_0|) \subset X$ and $f^{-1}X' \cap |\operatorname{St}^2 K_0|$ is a finite complex.

We will prove that we can give arbitrarily fine equivariant approximations f' to f such that f' coincides with f outside X, and f' is piecewise linear in K_0. In this extension business the Extension Lemma proves useful.

Lemma 92 (Modification of Illman [7], Lemma 2.6) *Let $f\colon |K, L| \to (X, X_0)$ be a G-equivariant C^r-map, and K/G a finite complex. Let, furthermore, S be an invariant subcomplex of K. Then, given $\delta > 0$, there exists δ_1 such that any (strong) δ_1-approximation $h|\colon (S, S \cap L) \to (X, X_0)$ to $f|$ can be extended into a (strong) δ-approximation $h\colon |K', L'| \to (X, X_0)$ such that h coincides with f and K' coincides with K outside $\overset{\circ}{St} S$.*

Proof: Since K/G is finite, there is only a finite number of equivariant simplices. Let the dimension of K be m, and denote the i-skeleton of K by K^i.

First we define $\delta_m = \delta$, and then define $\delta_{m-1}, \delta_{m-2}, \ldots, \delta_1$, $\delta_i < \delta$ for all i, inductively, in the order of decreasing dimension so that any (strong) δ_{i-1}-approximation h^{i-1} to

$$f\colon (K^{i-1} \cup S \cup (K \setminus \overset{\circ}{\operatorname{St}}\, S), (K^{i-1} \cup S \cup (K \setminus \overset{\circ}{\operatorname{St}}\, S)) \cap L) \to (X, X_0)$$

can be extended into a (strong) δ^i-approximation

$$h^i\colon (K^i \cup S \cup (K \setminus \overset{\circ}{\operatorname{St}}\, S), (K^i \cup S \cup (K \setminus \overset{\circ}{\operatorname{St}}\, S)) \cap L) \to (X, X_0)$$

to $f|$. This is possible by Corollary 91 and the fact that each skeleton has only a finite number of simplices so that we can apply the Corollary 91 in each simplex and take the minimum of δ's that it provides us with. In the case of a boundary neighbourhood, f is coplanar by the definition of a map, and by the last part of Lemma 90, h^i's are strong approximations and coplanar.

The subdivisions of the sets K^i that are the domains of the maps h^i are provided by the Corollary 150.

Now we can just choose $\delta_1 = \delta^1$, since if we are given the $h|S$, we can extend it inductively to simplexes in $(\overset{\circ}{\operatorname{St}}\, S)\setminus S$ in the order of increasing dimension with the sequence of δ^i's described above. □

Corollary 93 *The previous lemma works, if we replace (X, X_0) with (M, N), assume that $f(|\operatorname{St}^2 S|) \subset (X, X_0)$, where (X, X_0) is a coordinate neighbourhood pair for some point of X and speak about $(\cdot, \operatorname{St} S, (X, X_0))$-approximations.*

Proof: Apply Lemma 92 to $(\operatorname{St}^2 S, L \cap \operatorname{St}^2 S)$ as (K, L) and $f|\operatorname{St}^2 S$.

Changes in Lemma 92 are made only inside $\overset{\circ}{\operatorname{St}} S$. That we get the complex K' follows by applying Proposition 147 to K and the subdivision obtained from Lemma 92. (Notice that $|\partial \operatorname{St} S| \cap \overset{\circ}{\operatorname{St}} S = \emptyset$. This is seen as follows: If $x \in |\partial \operatorname{St} S|$, then $x \in |s|$, s is not a simplex of $\operatorname{St} S$. Then s and S do not intersect. However, if $x \in \overset{\circ}{\operatorname{St}} S$, then each simplex s' such that $x \in |s'|$ intersects S.)

Now $f'|\,|\operatorname{St}^2 S|$ (obtained from Lemma 92) and $f|\,|K \setminus \operatorname{St} S|$ coincide in the overlapping part, and thus we get the required f' by $f'(x) = f'|(x)$, if $x \in |\operatorname{St}^2 S|$, and $f'(x) = f(x)$ otherwise. □

Corollary 94 *Let (K, L) be a simplicial G-complex pair and $A \subset K$ an invariant subcomplex. Let $t\colon (K, L) \to (M, N)$ be an equivariant C^r-embedding, and $A' \subset A$ an invariant subcomplex such that $t \operatorname{St}^3 A' \subset X$ for a coordinate neighbourhood X. Let $\epsilon > 0$. Then there exists $\delta > 0$ such that any equivariant (δ, A', X) approximation $t'|$ to $t|A$ can be extended into a G-equivariant $(\epsilon, \operatorname{St}_K A', X)$-approximation to t.*

Proof: Let $B = (\operatorname{St} A') \cup A$. Enumerate the simplices of $B \setminus A$. Assume that $t'|A = t|A$ and proceed as in the proof of Lemma 92 to obtain a subdivision B' of B and $(X, \epsilon, \operatorname{St}_K A)$-approximation $t'|\colon (B', L \cap B' \to (M, N)$ to $t|$ such that B and B', and $t'|$ and $t|$ coincide in $A \cup (B' \setminus \overset{\circ}{\operatorname{St}} B')$.

Since changes appear only inside $\overset{\circ}{\operatorname{St}} A'$, the final $t'\colon K' \to M$ can be constructed by defining $t'(x) = t'|(x)$, if x in L, and $t'(x) = t(x)$ otherwise. We get K' in the same way as in the previous corollary. □

Now we know that we can extend approximations. Next we prove that we can extend approximations that are simplexwise linear.

Theorem 95 *Let $f\colon |K, L| \to (X, X_0)$ be a G-equivariant map and K/G a finite complex. Let, furthermore, S be a G-invariant subcomplex. Then, given $\epsilon > 0$, there exists an ϵ-approximation $h\colon |K', L'| \to (X, X_0)$ to f, such that $h|S'$ is linear in each simplex, and h coincides f and K' coincides K outside $\overset{\circ}{St} S'$.*

Furthermore, in the boundary case, we can assume that f' is a strong approximation to f.

Proof: Use Lemma 92 to obtain δ and then Corollary 85 to obtain a linear δ-approximation to $f|S$.□

Theorem 96 *Let (M, N) be a G-manifold pair, and let $f: |K, L| \to (M, N)$ be a G-equivariant C^r-map, K is locally finite. Let, furthermore, S be an invariant subcomplex of K. Assume, furthermore, that S/G is finite and $f \operatorname{St}^2 S \subset X$, where (X, X_0) is a coordinate neighbourhood pair. Then, given $\delta > 0$, there exists a $(\delta, \operatorname{St} S, X)$-approximation h to f such that $h|S$ is linear.*

Proof: In the previous theorem, use $(\operatorname{St} S, (\operatorname{St} S) \cap L)$ as (K, L) and obtain $f'|: ((\operatorname{St} S)', L \cap (\operatorname{St} S)') \to (X, X_0)$. Define (K', L') by replacing $\operatorname{St} S$ by $(\operatorname{St} S)'$ in K (see Proposition 148 in the Appendix), and define $f': (K', L') \to (M, N)$ so that it coincides with $f'|$ in S' and with f elsewhere. This gives us a well-defined equivariant simplicial complex (K', L'), and a C^r-equivariant map, since the simplicial structures of the overlapping part $(K \setminus \overset{\circ}{\operatorname{St}} S) \cap (\operatorname{St} S)'$ are the same in $(\operatorname{St} S)'$ and $K \setminus \overset{\circ}{\operatorname{St}} S$, and also $f'|((K \setminus \overset{\circ}{\operatorname{St}} S) \cap \operatorname{St} S')$ equals $f|((K \setminus \overset{\circ}{\operatorname{St}} S) \cap \operatorname{St} S')$.□

The rest of this section deals with using the simplexwise linear approximation and extension results in a case where we want to make approximations of two embeddings intersect in a subcomplex.

Theorem 97 *Let $f: |K_1, L_1| \to (M, N)$ and $h: |K_2, L_2| \to (M, N)$ be G-equivariant C^r-embeddings such that $f \operatorname{St}^3 f^{-1}(\operatorname{im} f \cap \operatorname{im} h), h \operatorname{St}^3 h^{-1}(\operatorname{im} f \cap \operatorname{im} h) \subset (X, X_0)$, where (X, X_0) is a coordinate neighbourhood. Assume, moreover, that each component of (X, X_0) intersects only a finite number of simplices. Then, given δ, there exist (strong) $(\delta, \operatorname{St}^2 f^{-1}(\operatorname{im} f \cap \operatorname{im} h), X)$ and $(\delta, \operatorname{St}^2 h^{-1}(\operatorname{im} f \cap \operatorname{im} h), X)$-approximations to f and h, respectively, which intersect in a full subcomplex.*

Proof: First we show the absolute case, that is, the case where L_1, L_2 and N are empty.

The distance of hK_2 from $f(K_1 \setminus f^{-1} \operatorname{St}(\operatorname{im} f \cap \operatorname{im} h))$ is positive, and also the distance of fK_1 from $h(K_2 \setminus h^{-1} \operatorname{St}(\operatorname{im} f \cap \operatorname{im} h))$ is positive. Thus we can assume that 2δ is smaller than these distances. Furthermore, let δ be so small that any δ-approximation to f and h is an embedding, and in the case of a boundary neighbourhood a strong approximation.

Use Theorem 96 to take δ-approximations f' and h' to f and h, respectively, so that they are linear in $\operatorname{St} f^{-1}(fK_1 \cap hK_2)$ and $\operatorname{St} h^{-1}(fK_1 \cap hK_2)$. By the choice of δ, $\operatorname{im} f'$ and $\operatorname{im} h'$ intersect only in the part where the maps were made linear. Thus we can, by subdividing further the linear part, make the simplicial maps intersect in a full subcomplex (possible by Theorem 77, see also the remark after the Theorem 77).

Since this subdivision only needs be made inside $\operatorname{St} f^{-1}(fK_1 \cap hK_2)$ and $\operatorname{St} h^{-1}(fK_1 \cap hK_2)$, the extended subdivision affects only inside the second-iterated stars. Thus, changes are made only inside $\operatorname{St}^2 f^{-1}(\operatorname{im} f \cap \operatorname{im} h)$ and $\operatorname{St}^2 h^{-1}(\operatorname{im} f \cap \operatorname{im} h)$.

Now the relative case is totally similar, since Theorem 96 was proved for pairs, and subdividing is always compatible with the pair structure. □

The previous theorem is what we were aiming at for further use in our existence result for triangulations. The uniqueness result requires a somewhat stronger result, which we prove next.

Corollary 98 (Similar to Munkres [18], Lemma 10.2) *Let $t\colon |K,L| \to (M,N)$ be C^r and G-equivariant, and let A and B be G-invariant subcomplexes of K such that $t|A$ and $t|B$ are closed embeddings, and $\mathrm{St}^4\, t^{-1}tA \subset (X,X_0)$, where (X,X_0) is a coordinate neighbourhood. Then, for every $\delta > 0$, there exists $t_1\colon |K_1', L_1'| \to (M,N)$ that is a (strong) $(\delta, \mathrm{St}^3\, t^{-1}tA, (X,X_0))$-approximation to t and such that t_1A and t_1B intersect in a full subcomplex.*

We, furthermore, have that $t_1|A\cup B$ is an imbedding, when we identify the overlapping simplices in $A \cup B$.

Proof: First we note that $(\mathrm{St}^4\, t^{-1}tA)\cap(A\cup B)$ has only a finite number of equivariant simplices, since that complex is mapped inside the componentwise relatively compact set X by a closed map.

We can assume that $K = (\mathrm{St}^2\, t^{-1}tA) \cap (B \cup A)$, since after we have a proof for this special case, we can extend the approximation to entire (K,L) by Lemma 92. For small enough δ, this extension does not cancel the required intersection condition of A and B, since the distance of tA and $(tB \setminus (\mathring{\mathrm{St}}_2 t^{-1}tA)) \cap X$ is positive. This follows from the fact that $t|B$ is a closed map.

Thus, as in the proof of the previous proposition, we take a linear δ-approximation t_1 to t such that $t_1|A$ and $t_1|B$ are closed imbeddings. Now as in the previous theorem, we can subdivide A' and B' (the parts of the domain of t_1 that correspond to A and B) as independent complexes, to obtain maps $t_1|A'$ and $t_1|B'$ that intersect in a full subcomplex. We do this subdivision without changing notation.

Now, we see that the subdivisions A' and B' give a subdivision of $K' = A' \cup B'$. Namely, let s be a simplex of A' or B' such that $|s| \subset |A \cap B|$. Then $t_1||A'|\,|s| = t_1||B'|\,|s|$, and because of the intersecting in a subcomplex property, s is a simplex of both A' and B'.

Now $t_1|(A\cup B)$, where we identify overlapping simplices, is an embedding by Lemma 80.□

Acknowledgements

This section is simply about using techniques of Munkres [18] and Illman [7] in our setting.

Chapter 4

Main results

Here we prove the existence and uniqueness results for triangulation. In the existence result, we need to piece together triangulations of parts of the manifold. So we start with a section where we prove that this piecing together works as expected.

4.1 Preliminary for the existence result: Glueing complexes together

Definition 99 *Let $t_i \colon |K_i| \to M$, $i \in I$, be embeddings of simplicial complexes into a space, and $L'_j \subset K_j$ a subcomplex for a each $j \in I$. We define that $\mathrm{St}_{i\in I}(L'_i)_{i\in I}$ is the collection $(L_i)_{i\in I}$ of subcomplexes of K_i, $L_i \subset K_i$ such that a simplex s is in L_i if and only if $t_i|s| \cap t_j|L'_j| \neq \emptyset$ for some $j \in I$.*

If $U \subset M$, then $\mathrm{St}_{i\in I}(L'_i)_{i\in I} \subset U$ means that $t_i L_i \subset U$ for all indices i.

We define also iterated stars by induction, that is, $St^n_{i\in I}(L'_i)_{i\in I} = St_{i\in I}St^{n-1}_{i\in I}(L'_i)_{i\in I}$ In other words, if $s \in K_i$ then $s \in (\mathrm{St}^n_I(L'_i)_{i\in I})_i$ if and only if there is a sequence of simplices $s_0, \ldots s_n = s$ of the complexes (K_i) such that the images of two adjacent simplices intersect and $s_0 \in L'_j$ for some $j \in I$.

We usually use this device for only one set L'_{i0} instead of the collection $(L'_i)_{i\in I}$: If $L'_{i0} \subset K_{i0}$, then $St^n_{i\in I}L'_{i0}$ means the collection $St^n_{i\in I}(L'_i)_{i\in I}$, as we define the sets L'_i as follows: L'_{i0} is as above, and $L'_i = \emptyset$ otherwise.

Definition 100 *If $t_i \colon K_i \to M$, $i \in I$ are injective maps that intersect pairwise in a full subcomplex, then we define a simplicial complex $K = \coprod/K_i$ and $t = \coprod/t_i$ as follows:*

- *$K = (\coprod K_i)/R$, where vRv' if and only if $v \in K_i$, $v' \in K_j$, $t_i(v) = t_j(v')$ for some i, j.*
- *If s is a simplex of K_i, then the image of s in the natural embedding $\pi_i \colon K_i \to \coprod/K_i$ is a simplex of $\coprod/K_i$; all simplices of $\coprod/K_i$ are of this type.*

- $t(x) = t_i(x)$ *for some* i *such that* $x \in |K_i|$.

If we have the relative case, namely $t_i\colon (K_i, L_i) \to (M, N)$, *we let* $\coprod/(K_i, L_i) = \coprod/(K_i, \bigcup_{i\in I} \pi_i L_i)$, *where* π_i *is the natural embedding of* K_i *into* $\coprod/K_i$.

Remark 101 *The complex* $K = \coprod/K_i$ *is well-defined if the maps* t_i *intersect in a subcomplex (and not necessarily in a full subcomplex.) If the maps* t_i *do not intersect in a full subcomplex, the map* $t = \coprod/t_i$ *need not be well-defined (i.e.,* $t(x) = t_i(x)$ *does not have to be independent of* i*.) We define* $K = \coprod/K_i$ *and* $t = \coprod/t_i$ *in all cases where the injective maps* t_i *intersect pairwise in a subcomplex, and the map* t *is well-defined.*

Lemma 102 *We have the following.*

1. $K = \coprod/K_i$ *is a well-defined simplicial complex, and if the maps* t_i *intersect pairwise in a full subcomplex,* $t = \coprod/t_i$ *is a well-defined simplicial map* $K \to M$.
2. *If* K' *is a subdivision of* K *and* $t'\colon K' \to M$, *then we have subdivisions* K_i' *of* K_i *and maps* $t_i'\colon K_i' \to M$ *such that* $K' = \coprod/K_i'$ *and* $t' = \coprod/t_i'$.
3. *If* $L_i' \subset K_i$, $i \in I$, *then* $\mathrm{St}_{\coprod/K_i} \bigcup_I L_i' = \coprod/\mathrm{St}_I(L_i')_{i\in I}$.
4. *If* t' *is a* $(\delta, X, \coprod/L_i)$*-approximation to* t, *then each* t_i' *is a* $(\delta, X, \coprod L_i \cap K_i')$ *- approximation to* t_i.
5. *If all* K_i*'s are* G*-complexes,* M *is a* G*-space, and all the maps* t_i *are* G*-equivariant, then* $\coprod/K_i$ *has a natural action of* G *such that the map* t *is* G*-equivariant.*
6. *The map* t *is injective.*

Proof: We prove (1). K is obviously a well-defined simplicial complex. Let $x = \sum \ell_j[v_j] \in K$. Assume $t_i(x) \neq t_{i'}(x)$. Then there exists a minimal simplex $s \in K_i$ such that $\sum \ell_j v_j \in |s|$. Now, for each vertex v of s we have that $t_i(v) \in t_{i'}|K_{i'}|$. Since K_i and $K_{i'}$ intersect in a *full* subcomplex, we have that $t_i|s| \subset t_{i'}|K_{i'}|$. Furthermore, $s' = t_{i'}^{-1} t_i s$ is a simplex of $K_{i'}$, and each $v_j' = t_{i'}^{-1} t_i v_j$ is a vertex of s'. Thus $t_{i'}x = t_{i'} \sum \ell_j v_j' = \sum \ell_j v_j = t_i x$, since the map $t_{i'}^{-1} t_i|s$ is linear.

(2) Let $s\colon |K'| \to |K|$ be the subdivision map. Each natural embedding $\pi_i\colon |K_i| \to |K|$ is injective. Let K_i' be the subcomplex of K' covered by the map $s^{-1} \circ \pi_i$ and $\pi_i^{-1} \circ s||K_i'|$ be the subdivision map $s_i\colon |K_i'| \to |K|$. Let $t_i'\colon |K_i| \to M$ be the map $t_i \circ s_i$. It is obvious that these choices work.

We prove (3). Let s be a simplex in $\mathrm{St}_{\coprod/K_i} \bigcup_i L_i'$. Then ts intersects ts', s' a simplex of L_j' for some j. But then $t_i s$ intersects $t_j s'$, and $s \in \coprod/\mathrm{St}_I(L_i')_{i\in I}$.

Let s be a simplex in $\mathrm{St}_I(L_i')_{i\in I}$. Then $t_i s$ intersects $t_j s'$ for some $s' \in L_j'$. But then $t_i s$ and $t_j s'$ have a common face $t_i s'' = t_j s'''$, and s'' and s''' are identified in $\coprod/K_i$. Thus $s \in \mathrm{St}_{\coprod/K_i} \bigcup_I L_i'$.

(4) is true, since passing to a subcomplex only relaxes the property of being an approximation.

We prove (5). We define an action of G in K such that $g[v] = [gv]$ for each vertex $[v]$ of K and each $g \in G$. We prove that the action is well defined. Let $[v] = [v']$. Then $t_i v = t_{i'} v$, and since the maps t_i, $t_{i'}$ are equivariant, we have that $t_i gv = t_{i'} gv'$. Thus $[gv] = [gv']$, and the action does not depend on the choices of representatives. That the action is simplicial follows from the definition of simplices of K. We have that $gt(x) = gt_i(x) = t_i(gx) = t(gx)$, for all $x \in |K|$ and $g \in G$. Thus the map t is equivariant.

We prove (6). Since $t||K_i| = t_i$ is injective for each i, we must prove only that for $x \in |K_i|$, $y \notin |K_i|$ we have that $t(x) \neq t(y)$. Assume in contrary that $t(x) = t(y)$. Then $t_i(x) = t_{i'}(y)$ for some i', and thus there exist simplices s of K_i and s' of $K_{i'}$ such that $x \in |s|$, $y \in |s'|$, and $s = s' \in K$. Thus $y \in |K_i|$, a contradiction. □

Lemma 103 *Let X be a coordinate space. Let $t_1 \colon K_1 \to X$ and $t_2 \colon K_2 \to X$ be G-equivariant embeddings which intersect in a subcomplex. If $t_1 | \operatorname{St} t_1^{-1}(t_2 K_2 \cap X')$ and $t_2 | \operatorname{St} t_2^{-1}(t_1 K_1 \cap X')$ are linear in each simplex, then $\coprod / t_i$ is an embedding.*

Proof: The map $\coprod / t_i$ is injective by the previous lemma. We have to prove that it is an immersion. The only place where it could fail is a point $x \in \coprod / K_i$ such that $\coprod / t_i x \in t_1 K_1$ and $\coprod / t_i x \in t_2 K_2$. By equivariance, we may assume that $t_1 x \in X'$.

But for such a point x there exists a neighbourhood $\operatorname{St} x$ in $\coprod / K_i$ such that $\coprod / t_i | \operatorname{St} x$ is linear in each simplex and thus an immersion by Lemma 80. □

Corollary 104 *Let t'_1, t'_2 be the approximations of f and h, respectively, obtained in Theorem 97. By reading through the proof, we observe that these satisfy the assumptions of the previous lemma in the coordinates of (X, X_0) of Theorem 97. Thus, by the previous lemma, $\coprod / t'_i$ is an embedding.*

Now we have all the properties of $\coprod /$ that we need for the existence result. Before proceeding, we prove one result that we will need in the appendix.

Proposition 105 *Let $K = \coprod / K_i$, where each K_i is a G-complex, and let $t_i \colon |K'_i| \to |K_i|$ be equivariant subdivisions that intersect pairwise in a subcomplex. Then $K' = \coprod / K_i$ is a well-defined G-complex, and $\coprod / t_i \colon |K'| \to |K|$ is an equivariant subdivision.*

Proof: Here we do not assume that the complexes K'_i intersect in a full subcomplex. However, K' can be defined as before. Now, the question is whether the map $\coprod / t_i$ can be defined. The answer in our case is yes. Assume $x \in |K'|$, $x \in |K'_i|$, $x \in |K'_{i'}|$. Let s be a minimal simplex of $|K'_i|$ containing x, and let s' be a minimal simplex of $|K'_{i'}|$ containing x. We have that for all vertices v of s, $t_i(v) = t_{i'}(v')$, where v' is a vertex of s' and vice versa (reversing the roles of i and i'). Since the maps $t_i||s|$ and $t_{i'}||s|$ are linear, $t_i x = t_{i'} x$.

Thus the map $\coprod/t_i$ exists. By Lemma 102 (5), we have that K' is a G-complex, and $\coprod/t_i\colon |K'| \to |K|$ is a simplexwise linear G-map.

The map $\coprod/t_i$ is trivially surjective. That it is injective follows from Lemma 102 (6).

□

Remark 106 *We remark that all the results stated in this section are true also for the* $\coprod/$ *of pairs.*

Acknowledgements

These results were used more or less implicitly by Munkres [18] and Illman [7]. Writing them down properly is my own contribution.

4.2 Existence of triangulation

We construct an equivariant C^r triangulation of a C^r manifold pair (M, N), $1 \leq r \leq \omega$. We assume that (M, N) is equipped with a properly discontinuous C^r-action of G. We, furthermore ,assume that $\partial N = N \cap \partial M$ and that ∂M and N are transversal on ∂N.

We choose a locally finite G-covering (U_i) by coordinate neighbourhoods of our manifold pair (M, N) and another covering (V_i) by coordinate neighbourhoods such that $V_i \subset\subset U_i$ for all indices i. These coverings exist by Theorem 40, and they are star-finite.

Next, we choose a simplicial G-complex K_i and an equivariant embedding $t_i\colon |K_i| \to U_i$ such that $V_i \subset\subset \operatorname{im} t_i$ for all indices i. This is possible by Lemma 74. For pairs, we furthermore require that each $t_i^{-1}N$ is a subcomplex of K_i. This can be achieved by the remark after Lemma 74.

Next, we will, in an inductive process, make small corrections to all imbeddings t_i defined at the beginning of this section so that their cellular structures coincide in the overlapping parts. Almost all the work was already done in the last chapter, and all we have to do is to be cautious that in the process every K_i gets subdivided only a finite number of times. This can, however, be done, since we can choose approximations so small that in every step the "overlap structure" of $K_i's$ is preserved.

We assume that our manifolds are second countable, and thus we can assume that the set of indices i is $\mathbb{N}$.

Inductive Step 107 (for index k) *We have simplicial G-complexes K_i^k and G-equivariant C^r-embeddings $t_i^k\colon K_i^k \to (M, N)$, $i \in \mathbb{N}$. We denote $t_i^k = t_i$ and $K_i^k = K_i$ for all $i \in \mathbb{N}$. The following hold:*

1. *For all $i \in \mathbb{N}$ and for all simplices s of $\operatorname{St}^5_{\mathbb{N}} K_i$, we have that $t_j|s| \subset U_i$, where j is such an index that s is a simplex of K_j.*

2. *For all $i, j \leq k$, K_i and K_j intersect in a full subcomplex.*

3. *For all $i \in \mathbb{N}$, we have that t_i is an equivariant embedding, and that $V_i \subset\subset t_i K_i$. In the case of pairs we, furthermore, require that $V_i \cap N \subset\subset t_i^k (t_i^0)^{-1} N$ (where $\subset\subset$ is taken in the relative topology of N.)*

4. *For all $i, j \in \mathbb{N}$, we have that $t_i^k K_i^k$ and $\bar{U}_j$ intersect only if $t_i^{k-1} K_i^{k-1}$ and $\bar{U}_j$ intersect.*

5. *For all $i \in \mathbb{N}$, we have that t_i^k and K_i^k coincide with t_i^{k-1} and K_i^{k-1} everywhere outside $(t_i^{k-1})^{-1} U_k$, that is, changes to the maps, and also subdivision is done only inside U_k. Furthermore, the maps t_i^k are changed only inside $St^3_{\mathbb{N}} K_k^k$.*

6. *The imbeddings $t_i^k \colon K_i^k \to (M, N)$ are compatible with the pair structure, that is, $(t_i^0)^{-1} N$ consists of entire simplices, and $t_i^k(s) \subset N$ if $|t_i^0|s \subset N$ for all i and all $s \in K_i$.*

7. *$\coprod_{i \leq k} / t_i$ is an embedding.*

8. *The maps t_i^k are coplanar and strong approximations to the maps t_i^{k-1}.*

Proposition 108 *The inductive process can be started.*

Proof: The complexes K_i and maps t_i as constructed at the beginning of this section satisfy (3). Conditions (2), (4), (5) and (7) are not required at the first step, since they make sense only at later rounds.

The point (1) can be made to hold by subdividing enough. Start by fixing an equivariant metric for (M, N). Let $i \in \mathbb{N}$ be fixed. The distance of $t_i|K_i|$ from the boundary of U_i is positive, let it be d. Now, since the diameter of simplices converges to 0 in the iterated barycentric subdivision (although the complexes are not finite, they are finite at the orbit space level, and that is good enough for us, since everything is equivariant), we can subdivide each K_j such that $t_j|K_j|$ and U_i intersect, until the following holds: If s is a simplex of K_j such that $B(t_i|K_i|, 5d/6)$ and $t_j|s|$ intersect, then $t_j|s|$ is contained in U_i, and the diameter of $t_j|s|$ is less than $d/6$. Now (1) is satisfied for the index i. This is due to the fact that all "chains" of successively intersecting 5 simplexes intersecting $t_i|K_i|$ have a diameter less than d.

We can repeat the process above for each $i \in I$. Since the covering U_i is star-finite, each simplex gets subdivided only a finite number of times. Thus (1) is satisfied.

(6) Follows from the remark after Lemma 74.

(8) Holds, since a simplexwise linear map is always coplanar. □

Proposition 109 *The inductive process can be continued, that is, if there exist complexes K_i^k and maps t_i^k, $i \in \mathbb{N}$ satisfying the points at the inductive step for k, then we can find the complexes K_i^{k+1} and maps t_i^{k+1}, $i \in \mathbb{N}$.*

Proof: We regard the complexes K_i^k and embeddings t_i^k, $i \leq k$, as one complex $K = \coprod / K_i^k$ and embedding $t = \coprod / t_i^k$, where we identify all overlapping simplices. Let $K'' = \operatorname{St} t^{-1}(tK \cap t_{k+1}^k K_{k+1}^k)$.

Then we choose δ so small that when we take the (δ, U_{k+1}, K'')-approximation t' to t, the corresponding "components" t_i' of t' satisfy all the points in the induction with the choice

- $t_i^{k+1} = t_i', i \leq k$,
- t_{k+1}^{k+1} is a $(\delta, U_{k+1}, K_{k+1}^k)$-approximation to t_{k+1}^k and
- $t_i^{k+1} = t_i^k, i > k+1$

except for (2), (6) and the last part and (5), which we handle later in this proof.

That we can choose δ small enough follows from: (All distances are measured in the Euclidean metric of U_{k+1}).

- (1) If s and s' are equivariant simplices (of possibly different K_j's) that both intersect U_{k+1} and one of them is contained in U_{k+1}, but $t_j s \cap t_{j'} s' = \emptyset$, then the distance of $t_j s$ and $t_{j'} s'$ is positive. Since there are only finitely many such equivariant simplices, we can choose δ so small that in $(2\delta, K'', U_{k+1},)$-approximations, $t_j' s$ and $t_{j'}' s'$ never intersect if they do not intersect in the original map. Now choose δ still to be so small that every (δ, K'', U_{k+1})-approximation of t keeps $\operatorname{St}_{\mathbb{N}}^5 K_j$ (a finite number of equivariant simplices) inside U_j, for every j such that $U_j \cap U_{k+1} \neq \emptyset$.

- (3) Let for each $j < k$ such that K_j^k and $\bar{U}_{k+1}$ intersect $V_j' = V_j \cap \operatorname{St}_{\mathbb{N}}^3 K''$. That we can choose δ as required follows from Proposition 56, as we choose A to be V_j'. The rest of the sets V_j are covered by the sets $\operatorname{im} t_j^{k+1}$, since we have the requirement that we make no changes to the map t_j outside $\operatorname{St}_{\mathbb{N}}^3 K_k$. The point about pairs can be made to hold (provided that (6) holds) in a similar fashion by using Proposition 56 with $V_j' \cap N$ as A, N as the basic manifold, $N \cap U_k$ as the coordinate structure, and $(t_i^0)^{-1} N \subset K_j^k$ as the simplicial complex.

  We repeat this for all j such that $\bar{U}_k$ and K_j^k intersect, altogether only a finite number of times. When we take the minumum of the δ's, it is positive.

- (4) Since, by 5, changes are made only inside a componentwise compact set lying in U_{k+1}, and the distance of K_i^k and $\bar{U}_j$ is positive if they do not intersect, it is obvious that this can be made to hold.

- (5) This is obvious from the fact that we are dealing with a (δ, K'', U_{k+1})-approximation, and the inductive assumption (1).

- (8) By Proposition 70, an accurate enough approximation preserves coplanarity and makes any accurate enough approximation a strong approximation.

Since $K'' = \operatorname{St} t^{-1}(\operatorname{im} K \cap \operatorname{im} K_{k+1})$, the existence of a (δ, K'', U_{k+1})-approximation to t and a $(\delta, K_{k+1}^k, U_{k+1})$-approximation to t_{k+1}^k required to satisfy (2) and (6) follows from Theorem 97 and the inductive assumption (1). That (7) is satisfied follows from Corollary 104. □

Assume now that the G-complexes K_i' and the G-equivariant embeddings t_i', $i \in \mathbb{N}$, are the maps that are the result of the inductive process. Since the covering $\{U_i\}$ is star-finite, and $\operatorname{im} t_i^k \cap U_j \neq \emptyset$ only if $\operatorname{im} t_i \cap U_j \neq \emptyset$ (Inductive assumption 4), each of the complexes K_i and maps t_i is changed only a finite number of times in the inductive process (Inductive assumption 5). Thus each K_i' is a simplicial complex (i.e., no simplex is subdivided an infinite number of times), and each t_i' is a closed embedding whose image covers $\bar{V}_i$ (Inductive assumption 3). Now the required triangulation (K', t') can be obtained from the family $(K_i', t_i')_i$ by identifying overlapping simplices, since by the inductive assumption (2), each pair t_i', t_j' intersect in a full subcomplex.

Thus we have (almost) proved

Theorem 110 *Each C^r-differentiable G-manifold pair (M, N), the action properly discontinuous and C^r, has a G-equivariant C^r-triangulation.*

Proof: The only remaining thing to show the absolute case, that is the case where $N = \emptyset$, is that the map t is an immersion. For, we have not shown that the angle of two different simplices from different K_i's is positive. But each $\coprod_{i \le k} / t_i^k$ is an embedding, and for each $x \in M$, $t'|V$ and $\coprod_{i \le k} t_i^k|V$ coincide for some $k \in \mathbb{N}$ and a neighbourhood V of x, and $\coprod_{i \le k} t_i^k|V$ is an immersion.

Consider the relative case, that is, the case where $N \neq \emptyset$. Now $L_i' = t_i'^{-1} N$ is a subcomplex of K_i', and by the inductive assumption (3), $t_i' L_i'$'s cover N. Thus, if we choose $(K', L') = \coprod / (K_i', L_i') = (\coprod / K_i', \bigcup_{i \in I}(L_i'))$, we have a triangulation $t' \colon (K', L') \to (M, N)$. □

Corollary 111 *The assumption that M is N_2 is unnecessary; the previous theorem is true for all paracompact manifolds.*

Proof: Each connected component of a paracompact manifold M is second countable. Thus we choose one connected component C of M from each orbit of the component space of M and triangulate it G_C-equivariantly to obtain $t_C \colon K_C \to C$. By Corollary 13 we have that $GC = G \times_{G_C} C$, and thus we get a triangulation $G \times_{G_C} t_C \colon |G \times_{G_C} K_C| \to GC = G \times_{G_C} C$. We repeat this for each chosen C and take the disjoint union of the obtained triangulations. Pairs do not produce any extra difficulties.□

Acknowledgements

The basic idea is from Munkres [18] and Illman [7], but we have made quite drastic changes. For differences in our strategies, see the remark at the end of Section 4.4.

Illman and Munkres use a theorem on dimension theory which allows them to deal with only a finitely many K_i, and the circumvention of this caused the induction to have more assumptions.

4.3 Preliminary for the uniqueness result: subdividing

For a map $|K| \to M$ to be regarded as a triangulation, it is not necessary that the image of each simplex lies in a coordinate neighbourhood. However, when proving the uniqueness up to subdivision and arbitrarily small changes, we must be able to operate inside coordinate neighbourhoods.

In this section, we show some elementary results that let us pass into a situation where we can operate inside coordinate neighbourhoods.

Lemma 112 *Let $|K|$ be a locally finite simplicial G-complex. If $t\colon |K| \to M$ is G-equivariant and C^r and (U_i) is a covering of M with coordinate neighbourhoods, then there exists a G-equivariant subdivision K' of K such that for each simplex s of K' we have that $t(s) \subset U_i$ for some i.*

Proof: It is enough to prove that for a non-equivariant, countable K and a locally finite covering $\{U_i\}$ of $|K|$, there exists a subdivision K' such that each simplex of K' is contained in some U_i. The full case follows by Lemma 81, as one applies the special case to the complex K/G and neighbourhoods $\pi t^{-1}\{U_i\}$, after taking a 2nd barycentric subdivision of K to ensure that the condition in Lemma 81 is satisfied. Namely, the subdivision of K/G gives a subdivision of K in a natural way. Countability can be assumed, since each connected component of K/G is countable, and different components can be subdivided separately.

We enumerate the maximal-dimensional simplexes of K and subdivide them in an inductive process so that after each step i, the ith simplex s_i of K is subdivided so that it satisfies the condition that if s' is a simplex of the subdivision of s_i, then $s' \subset U_i$ for some i, and after the ith step, some small amount of other simplexes in the star of s are also subdivided (see the standard extension in the proof of Corollary 78) so that after each stage we have a complex. Since the ith simplex may be subdivided before the step i, we may have to handle at the step i a complex consisting of several simplexes, but anyway only a finite complex.

Let then s be the simplex (or the finite complex that corresponds to the original simplex s) that we want to subdivide. Since $|s|$ is compact there is a Lebesque number n_s for the covering $t^{-1}(U_i) \cap s$ (give s any metric you like). Subdivide s, or the finite complex that corresponds to s, by iterated barycentric subdivisions until each new simplex has diameter $< n_s$. Extend the subdivision to the entire K by the standard extension (see Munkres [18], Lemma 7.12). This changes only the star of s. Now the step i is complete.

Since at each stage changes are made only to the star of s_i, each simplex gets subdivided only some finite number of times during the process, and thus the process as a whole gives us the required subdivision K'. □

Lemma 113 *Let $t_1\colon K_1 \to M$ and $t_2\colon K_2 \to M$ be G-equivariant C^r embeddings of simplicial complexes into a G-manifold, and let (U_i) be an open G-covering of M. Let, furthermore, $n \in \mathbb{N}$. There are G-equivariant subdivisions K_1' and K_2' of K_1 and K_2, respectively, such that for any simplex ℓ of K_2' the set $\mathrm{St}^n_{\{1,2\}}\,\ell \subset U_i$ for some i. (Here the star is taken with respect to K_1' and K_2'.)*

Proof: The orbit space of M is paracompact by Abels-Strantzalos, Proposition 3.2.5.D. Thus the covering (πU_i) has an n times iterated star-refinement V_i (see Pears [20] 2.1.27). Let (V_i') be a locally finite covering with coordinate neighbourhoods refining $(\pi^{-1}V_i)$. Let K_1' and K_2' be the subdivisions obtained from Lemma 112 w.r.t. the covering (V_i').

Let now ℓ be a simplex of K_2'. Let $t_2\ell \subset V_j$, and U_i be such that $\mathrm{St}^n\, V_j \subset U_i$. If now $\ell' \in \mathrm{St}^n_{\{1,2\}}\,\ell$, then $\ell' \subset U_i$.□

4.4 Uniqueness of triangulation

Here we prove that the triangulation is in a certain sense unique: If (M, N) is a C^r, G-manifold pair and $t_1\colon (K_1, L_1) \to (M, N)$ and $t_2\colon (K_2, L_2) \to (M, N)$ are its equivariant C^r-triangulations, then there are G-equivariant C^r-triangulations $t_1'\colon (K_1', L_1') \to (M, N)$ and $t_2'\colon (K_2', L_2') \to (M, N)$ such that ${t_2'}^{-1}t_2'$ is a G-equivariant simplexwise linear isomorphism $(K_2', L_2') \to (K_1', L_1')$, and the primed complexes are G-equivariant subdivisions of their non-primed equivalents.

We also want that t_1' and t_2' can be, in some sense, chosen to be arbitrarily close approximations to t_1 and t_2, respectively. The sense is as follows:

Definition 114 *Let $\{U_i\}_{i\in I}$ and $\{V_i\}_{i\in I}$ be locally finite coverings of M with coordinate neighbourhoods such that $V_i \subset\subset U_i$ for all indices i, and let $(\epsilon_i)_{i\in I}$ be a collection of positive reals. Then we say that $t_1'\colon |K_1'| \to M$ is an $(\epsilon_i)_{i\in I}$-approximation to $t_1\colon |K_1| \to M$, if for every $x \in |K_1|$ such that $t_1(x) \in V_i$ we have that $t_1'(x) \in U_i$, $|t_1(x) - t_1'(x)| < \epsilon_i$ and $|dt_x(\bar{v}) - dt_x'(\bar{v})| < \epsilon|\bar{v}|$ for all indices i and tangent vectors $\bar{v}$. Here the distances are measured by the Euclidean structure of U_i, and K_1' is an equivariant subdivision of K_1.*

For now we will assume that (M,N) is N_2. The proof of the theorem that two triangulations can be subdivided to be "arbitrarily close to each other" is done in an inductive process that is similar to the one we used when proving that the triangulation exists. Here, however, things are a little more complicated. As in the first case, we have here the construction so far (a common, already matched subcomplex), and the new simplex we want to add to the construction, but also we must, at each step, extend the embeddings of the common subcomplex into the entire triangulations.

Let $t_1\colon (K_1, {}^0K_1) \to (M,N)$ and $t_2\colon (K_2, {}^0K_2) \to (M,N)$ be triangulations.[1] Let $\ell_1, \ell_2, \ldots$ be an enumeration of equivariant simplices of a subdivision of K_2 (the subdivision will be defined at the start of the induction) in the order of increasing dimension (meaning that the faces of any simplex appear in the enumeration before the simplex itself), and let $L_i = \ell_1 \cup \cdots \cup \ell_i$ for all $i \in \mathbb{N}$.

In the inductive process we regard these initial systems as systems with the index -1.

Let $(U_i)_{i\in\mathbb{N}}$ be a locally finite covering of M with coordinate neighbourhoods, and $(V_i)_{i\in\mathbb{N}}$ a similar covering with coordinate neighbourhoods such that $V_i \subset\subset U_i$ for all $i \in \mathbb{N}$. (Recall that we assume that (M,N) is N_2.) These coverings can be found by Theorem 40. Let, furthermore, $(\epsilon_i)_{i\in\mathbb{N}}$ be a collection of positive reals.

We moreover assume that we have a function $\iota\colon \mathbb{N} \to \mathbb{N}$. This will be defined at the start of the induction.

Since we are constantly dealing with two indices, 1 and 2, we will use $\beth$ (the Hebrew letter "Beth") as a variable that gets these two values. For simplicity of notation, we leave out the universal quantification "for all values of $\beth$".

To each index i, we attach a subcomplex $V'^i_\beth$ of a subdivision of $K_\beth$ such that $V_i \subset\subset t_\beth V'^i_\beth \subset\subset U_i$ for all i. These subcomplexes and subdivisions of $K_\beth$ will be defined at the start of the induction.

Inductive Step 115 (for index k) *There exist G-equivariant C^r -embeddings $t^{k+1}_\beth\colon K^{k+1}_\beth \to (M,N)$ such that the following hold.*

1. *$t^{k+1}_\beth$ is a strong $(\epsilon_{\iota(k+1)}2^{-(k+1)}, A_\beth, U_{\iota(k+1)})$ -approximation to $t^k_\beth$ for some $A_\beth$. Furthermore, the distance of $t^k_\beth(x)$ and $t^{k+1}_\beth(x)$ is smaller than $2^{-(k+1)}\epsilon_i$, and $|D_x t^k_\beth \bar{v} - D_x t^{k+1}_\beth \bar{v}| \leq 2^{-(k+1)}\epsilon_i |\bar{v}|$ in the Euclidean metric of U_i for every i, every $x \in V'^i_\beth$ and every tangent vector $\bar{v}$ such that $t^k_\beth(x) \in U_i$.*

2. *For all $i \in \mathbb{N}$ the following holds: For each simplex $s \in \mathrm{St}^6_{\{}1,2\}\ell_i$, we have that $t^{k+1}_j|s| \subset U_{\iota(i)}$, where $j \in \{1,2\}$ is such an index that s is a simplex K^{k+1}_j. Here the star is taken with respect to $K^{k+1}_1 \cup K^{k+1}_2$. We remark that since we subdivide K_2 in the process, ℓ_i might be a subcomplex of K^{k+1}_2 with several equivariant simplices.*

[1] Sorry for the bad notation for subcomplexes, but I'm running out of symbols.

3. $t^{k+1}_{\beth}$ *is an equivariant embedding (and, as an embedding, coplanar), and* $V_i \subset\subset t^{k+1}_{\beth} V'^{i}_{\beth} \subset\subset U_i$ *for all indices* i *and* $\beth$. *In the case of pairs, we also require that* $V_i \cap N \subset\subset (t^{k+1}_{\beth} {t^{0}_{\beth}}^{-1} N) \cap V'^{i}_{\beth} \subset\subset U_i$.

4. K^{k+1}_1 *and* L^{k+1}_{k+1} *intersect in a full subcomplex.*

5. *For each simplex* s *of* $K_{\beth}$ *and each* U_i, $t^{k+1}_{\beth} s$ *and* $\bar{U}_i$ *intersect only if* $t^{k}_{\beth} s$ *and* $\bar{U}_i$ *intersect.*

6. *In the case of pairs,* $t^{k+1}_{\beth}$ *are compatible with the pair structure, that is,* $t^{k+1}_{\beth}(t^{0}_{\beth})^{-1} N \subset N$.

Lemma 116 *The inductive process can be started.*

Proof: Subdivide $K_{\beth}$ into $K^{0}_{\beth}$ using Lemma 113, such that for each simplex ℓ of $K^{0}_{\beth}$ we have that $\mathrm{St}^{6}_{\{1,2\}}\,\ell \subset U_i$ for some i. Enumerate the simplices of K_2 in such an order that the faces of any simplex precede the simplex itself. Let the simplices be $\ell_1, \ell_2, \ell_3, \ldots$. Choose ι such that $\mathrm{St}^{6}_{\{1,2\}}\,\ell_i \subset U_{\iota(i)}$. Let $t^{0}_{\beth} = t_{\beth}$. Thus (2) is satisfied. Choose an invariant metric for M and subdivide for all i until the diameter of simplices intersecting U_i is less than the distance of ∂V_i and ∂U_i. Choose $V'^{i}_{\beth} = \mathrm{St}\, t^{-1}_{\beth} \bar{V}_i$. This takes care of the first part of (3). The rest of (3), (1) and (5) are satisfied because of the choice $t^{0}_{\beth} = t_{\beth}$. (4) is vacuous at this stage of induction, and (6) is automatically satisfied. □

Lemma 117 *The inductive process can be carried on. That is, if the induction assumption is satisfied for* k, *it can be made to be satisfied for* $k+1$.

Proof: We operate inside $U_{\iota(k+1)} = U$. Let $\delta > 0$ (we will make more requirements for δ during the proof).

Let $K' = K^{k}_1 \cup \ell^{k}_{k+1}$, where we identify the faces of ℓ^{k}_{k+1} that belong to L^k to the corresponding simplices of K^{k}_1, and define $t' \colon K' \to (M, N)$ by $t' = t^{k}_1$ in K^{k}_1 and $t' = t^{k}_2$ in ℓ^{k}_{k+1}. By the inductive assumption (4) this gives us a simplicial complex and a simplicial map.

Use Corollary 98 and the inductive assumption (2) to obtain a $(\delta, \mathrm{St}^3\, t'^{-1} t' \ell_{k+1}, U)$-approximation t'' to t' such that K^{k+1}_1 and ℓ^{k+1}_{k+1} intersect in a full subcomplex, and t'' is an embedding after diminishing δ and identifying overlapping simplices. Now let $t^{k+1}_1 = t'' | K^{k}_1$ and K^{k+1}_1 be the suitable part of the domain of t''.

Let $\epsilon > 0$ (more requirements later) and diminish δ so that $\delta < \epsilon$ and δ is so small that $t'' | L^{k}_{k} \cup \ell^{k}_{k+1}$ can be extended into an $(\epsilon, \mathrm{St}^4_k\, \ell_{k+1}, U)$-approximation t^{k+1}_2 to t^{k}_2. This δ exists by Corollary 94.

With a suitable choice of ϵ, these will be the required $t^{k+1}_{\beth}$. Now (4) is satisfied. By Corollary 94 the point (6) can be made to be satisfied, by treating ${t^{0}_{\beth}}^{-1} N$ as L in the statement of Corollary 94.

Now (2), (3), (5) and the first part of (1) can be made to hold by choosing ϵ small enough, and these are similar to the corresponding points in the proof of Lemma 109.

The last part of (1) follows for small enough ϵ from Proposition 54 and (3), since the part of $t^k V'^i_\beth$ where we make changes is a subset of both U_i and U. □

Theorem 118 *Given G-equivariant C^r-triangulations $t_1\colon (K_1, {}^0K_1) \to (M,N)$, $t_2\colon (K_2, {}^0K_2) \to (M,N)$, similar locally finite open coverings $(V_i) \subset\subset (U_i)$ of M and a collection (ϵ_i), $\epsilon_i > 0$, there exist G-equivariant (ϵ_i)-approximations $t'_1\colon (K'_1, {}^0K'_1) \to (M,N)$ and $t'_2\colon (K'_2, {}^0K'_2) \to (M,N)$ to t_1 and t_2, respectively, that coincide, that is, they differ only in the naming of vertices. Furthermore, t'_1 and t'_2 are G-equivariant C^r-triangulations.*

Proof: Because in the inductive process changes are made only inside $U_{\iota(i)}$, and because of the condition (5) and the local finiteness of (U_i), each simplex of $K_\beth$ is changed in the process only a finite number of times. Thus we can define the end result complexes and the end result approximations of the process, and these are the required. They intersect in a subcomplex by (4), are triangulations by (3), (ϵ_i)-approximations by (1) and (3).

By (6) and the last part of (3), we have that ${}^0K_\beth = t'^{-1}_\beth N = t^{-1}_\beth N = {}^0K_\beth$. Thus the maps $t'_\beth$ are triangulations of pairs $(K'_\beth, {}^0K'_\beth) \to (M,N)$. □

Next we prove that $t_\beth$ are equivariantly homotopic to $t'_\beth$. In Munkres [18], this is seen as a part of the uniqueness of the triangulation, but in Illman [7] this homotopy property is absent.

Lemma 119 *If t is an (δ, L, X)-approximation to t', then t and t' are G-equivariantly C^r-homotopic. This holds, if $r \neq \omega$. If $r = \omega$, we have to confine ourselves to a C^∞-homotopy.*

Proof: $t, t'\colon |K| \to M$. By using the equivariant linear structure of the special component of X, we define

$$H\colon |K| \times \mathbb{R} \to M; H(k,x) = f(x)t(x) + (1 - f(x))t'(x).$$

Here f is an increasing smooth map $\mathbb{R} \to \mathbb{R}$ such that $f]-\infty, 0] = 0$ and $f[1, \infty[= 1$. See Lemma 24 for the definition of a linear combination. For points x not in $|L|$, $t(x)$ and $t'(x)$ coincide, and we can define $H(k,x) = t(x) = t'(x)$. Now H is the required homotopy; it is equivariant, since each $g \in G$ is linear in the coordinates of X. □

Theorem 120 *If $r \neq \omega$, then the maps $t_\beth$ are equivariantly C^r-homotopic to the maps $t'_\beth$. If $r = \omega$, then the maps $t_\beth$ are equivariantly C^∞-homotopic to the maps $t'_\beth$.*

Proof: From the previous lemma we get homotopies $H^k_\beth$ between $t^{k-1}_\beth$ and $t^k_\beth$, and by changing the coordinates of $\mathbb{R}$, we can assume that $H^k_\beth(x,t) = t^{k-1}_\beth(x)$ if $t \leq 1 - 2^{-(k-1)}$ and $H^k_\beth(x,t) = t^k_\beth(x)$, if $t \geq 1 - 2^{-k}$.

Now we can define the final homotopies as follows:
$H_\daleth(x,t) = H^k_\daleth(x,t)$, if $1-2^{-(k-1)} \leq t \leq 1-2^{-k}$,
and $H_\daleth(x,1) = t'_\daleth(x)$.

This homotopy is continuous and C^r, since f of the previous Lemma is C^r, and each point of $|K| \times \mathbb{R}$ has a neighbourhood where all but a finite number of $H^k_\daleth$'s are constant in the direction of the parameter t. $\square$

Next we drop the assumption that the manifold pair (M, N) is second countable.

Corollary 121 *In Theorems 118 and 120, the assumption about the manifold (M, N) being N_2 is unnecessary. The theorems are true for all paracompact manifold pairs.*

Proof: First we remark that it is enough to consider the case where $M = GC$ for a connected component C of M. The general case is a disjoint union of such special cases.

The only places in the proof where we used the N_2-property were indexing the open coverings (U_i) and (V_i) with natural numbers and indexing the equivariant simplices of K_2 with natural numbers. If $M = GC$, then every coordinate neighbourhood of M intersects C, and C is N_2. Thus the locally finite coverings (U_i) and (V_i) must be countable. Similarly, every equivariant simplex ℓ of K_2 is such that $t_2\ell \cap C \neq \emptyset$. Since $t_2|t_2^{-1}C$ is a triangulation of C, we have that K_2 can contain only countably many equivariant simplices.$\square$

Thus we have proved the following:

Theorem 122 *If $t_1\colon (K_1, {}^0K_1) \to (M, N)$ and $t_2\colon (K_2, {}^0K_2) \to (M, N)$ are G-equivariant C^r-triangulations of a paracompact C^r G-manifold pair (M, N), then there exist their arbitrarily close equivariant approximations t'_1 and t'_2, respectively, that are G-equivariant C^r-triangulations of (M, N) and intersect in a full subcomplex. Furthermore, t_1 is G-equivariantly homotopic to t'_1 and t_2 is G-equivariantly homotopic to t'_2. If $r \neq \omega$, then the homotopy is of the class C^r, and if $r = \omega$, then the homotopy is of the class C^∞.*

Acknowledgements

Here I use techniques of Munkres [18] and Illman [7], but due to the use of local coordinates in the definition of an approximation, I had to make some changes. The points (1), (2) and (5) in the inductive step 115 were added by me, intended to preserve the inductive process compatible with the local coordinate structure.

Remark

We remark that uniqueness of the triangulation in the equivariant category is a stricter property than uniqueness in the non-equivariant category. Thus our uniqueness result is *not* a special case of the corresponding uniqueness result in Munkres [18]. Namely, we deal with uniqueness up to arbitrarily small C^1-changes and subdivision. In the non-equivariant category one can make the changes and subdivision without considering the action at all. In the equivariant category one needs to make the C^1-changes and subdivision in an equivariant way.

Remark

Like Munkres [18] and Illman [7], we could use the uniqueness theorem to show that any two embeddings have arbitrarily fine approximations that intersect in a full subcomplex, and use this to prove the existence of triangulations. This could be done as follows: From dimension theory it follows that any covering of a manifold M has a refinement where only $\dim M + 1$ elements cover each point. This gives us (after a small amount of work) an equivariant cover of M with $\dim M + 1$ elements, each a disjoint union of subsets of $\mathbb{R}^n$. Then we could triangulate those pieces of $\mathbb{R}^n$ (or, to avoid infinity problems, their subsets so that they altogether cover M), and then in an inductive process of $m+1$ steps use a result similar to the uniqueness result, but formulated for embeddings instead of triangulations, to fit the $m+1$ partial triangulations together.

I have, however, chosen to prove the existence of triangulation separately. There are two reasons. First, I like to keep proofs as elementary as possible, and the reference to dimension theory in an otherwise elementary proof violated my sense of elegance.

Second, the inductive process is pretty complicated, and it took me months to grasp what actually happens in the induction where we fit complexes together. I hope that giving the simpler proof of existence first makes it easier to understand what actually happens.

4.5 The role of proper discontinuity

Now, one may ask whether a manifold can have an equivariant triangulation if the action is not properly discontinuous. We answer this question here by studying actions on simplicial complexes.

It is obvious that some manifolds M can be triangulated equivariantly even if the action is not properly discontinuous. Let, for example M, be equipped with the trivial action of $\mathbb{Z}$. Then, obviously, the action is not properly discontinuous, but M can be triangulated in an equivariant way.

However, the situation of our example is as bad as it can be, as we prove with a sequence of propositions.

We start by characterizing properly discontinuous actions in simplicial complexes.

Lemma 123 *Let K be a locally finite simplicial G-complex such that for each vertex $v \in K$ the set G_v is finite. Then the induced action in $|K|$ is properly discontinuous.*

Proof: Let $x, y \in |K|$, and let $U = \overset{\circ}{\mathrm{St}}\, x, V = \overset{\circ}{\mathrm{St}}\, y$ be neighbourhoods of x and y, respectively, that consist of a finite number of open simplices. If $gU \cap V = 0$, then there exist vertices $v \in \mathrm{St}\, y$ and $v' \in \mathrm{St}\, x$ such that $v = gv'$. For each pair v, v', we can have $gv = v'$ for only a finite number of elements g, and there are only finitely many choices of $v \in \mathrm{St}\, y$, $v' \in \mathrm{St}\, x$, thus $gU \cap V = 0$ for only finitely many g.□

It is easy to construct a simplicial complex with an effective action that is not properly discontinuous. We aim at showing that this is not possible in simplicial complexes that triangulate connected manifolds. Thus, we need to restrict the class of simplicial complexes we study. Slightly modifying Spanier [23] p. 148, we give the following definition.

Definition 124 *A locally finite simplicial complex K is an n-dimensional connected pseudomanifold if the following hold:*

1. *Each k-simplex, $k < n$, is a face of an n-simplex. There exist no simplices of dimension greater than n.*
2. *Each $n-1$-simplex of K is a face of at most two n-simplices of K.*
3. *Each pair s, s' of n-simplices of K can be joined with a finite chain $s = s_1, s_2, \ldots, s_m = s'$ of n-simplices of K so that any two adjacent simplices have a common $n-1$-face.*

It is known that a triangulation of a connected C^r-manifold M is a connected pseudomanifold K. Anyway, I could not find a reference, so I show it here.

Proposition 125 *Let $f\colon |K| \to M$ be a C^r-triangulation of a connected manifold. Then K is a connected pseudomanifold.*

We remark that in this propositon we do not have an action of G.

Proof: K is locally finite, since M is locally compact, and $|K|$ is locally compact (both in the metric and weak topologies) if and only if K is locally finite.

(1) Holds by the invariance of dimension.

(2) Let s be an $n-1$-simplex. Now $f|s|$ is an $n-1$-dimensional submanifold of M, and an interior point $f(x)$ of $f|s|$ has a coordinate neighbourhood (X, X_0), where $X \cap f|s| = X_0$ such that $X \subset f\overset{\circ}{\mathrm{St}}\, x$. Now, by the coordinate neighbourhood structure, removing $f|s|$ from X divides $X \setminus f|s|$ into 1 (if s is a boundary) or 2 connected components. But the number of connected components is at least the number of n-cells whose face s is.

(3) Let $A_1, A_2, \ldots$ be the equivalence classes of n-simplices when two simplices are equivalent if Condition 3 in the definition of a pseudomanifold is satisfied for them.

Assume that there are more than one equivalence class. Since M is connected, there must exist n-simplices s and s' in different classes which have no $n-1$-face in common but have a k-face t in common. Let t be of maximal dimension $< n-1$. Let (X, X_0) be a coordinate neighbourhood, $X \cap f|t| = X_0$, for a point $f(x)$ in the interior of $f|t|$ so that $X \subset \overset{\circ}{\mathrm{St}}\, f(x)$. Then, since t is of less dimension than $n-1$, removing $f|t|$ from X does keep $X \setminus f|t|$ connected. But from the simplicial structure we see that $f|s| \cap X$ and $f|s'| \cap X$ belong to different connected components of $X \setminus f|t|$, which is a contradiction. Thus K is a pseudomanifold.□

Proposition 126 *Let K be a connected simplicial G-complex that is a connected pseudomanifold. Let $H < G$ be the subgroup of those $g \in G$ for which $gv = v$ for every vertex $v \in K$. Then the induced action of G/H is such that $(G/H)_v$ is finite for every vertex v of K.*

Proof: From Abels-Strantzalos [1], Section 1.4, it follows that H is a normal subgroup and G/H actually is a group.

Let s be an arbitrary n-simplex of K. Let $g \in G$ be such that it keeps all the vertices of s fixed. Let s' be an n-simplex that has a common $n-1$-face with s. Then, by Condition 2 in the definition of a pseudomanifold, g must keep all the vertices of s' fixed, since s' has only one vertex that is not a vertex of s.

Using Condition 3 in the definition of a pseudomanifold and the fact stated in the previous paragraph iteratively, we see that g keeps all n-simplices of K fixed. By Condition 1 in the definition of a pseudomanifold g keeps every vertex of K fixed. Thus $g \in H$ if and only if g keeps one n-simplex pointwise fixed.

Let $v \in K$ be a vertex. We know that the action of G/H in K gives us an action of $(G/H)_v$ in $\mathrm{St}\, v$. If $g, g' \in (G/H)_v$ induce the same map $\mathrm{St}\, v \to \mathrm{St}\, v$, then $g^{-1}g'$ keeps an n-simplex pointwise fixed. Thus, by the argument in the previous paragraph, $g^{-1}g' = e_H$, where e_H the neutral element of H, and thus $g = g'$. Since K is locally finite, the complex $\mathrm{St}\, v$ is finite, and there are only a finite number of simplicial maps $\mathrm{St}\, v \to \mathrm{St}\, v$. Since different elements of $(G/H)_v$ correspond to different maps $\mathrm{St}\, v \to \mathrm{St}\, v$, the group $(G/H)_v$ is finite. □

Thus, we have proved

Theorem 127 *Let G act on a connected C^r-manifold M, and $H < G$ be the subgroup that keeps every point of M fixed. If there exists a G-equivariant triangulation of M, then the induced action of G/H on M is properly discontinuous.*

Proof: Let $t\colon |K| \to M$ be an equivariant triangulation. By Proposition 125, K is a pseudomanifold, and by Proposition 126, $(G/H)_v$ is finite for all the vertices v of K. By Lemma 123, the action of G/H on $|K|$ is properly discontinuous, and since t is an equivariant homeomorphism, the action of G/H on M is properly discontinuous.□

Acknowledgements

Except for the definition of a pseudomanifold and Proposition 125, this is totally my original work. The proof of Proposition 125 was worked out by myself, since I did not find a reference.

4.6 Putting the results together

The results in this thesis can be collected into the following theorem. In this theorem we also relax a little bit the assumption that the action is properly discontinuous.

Theorem 128 *Let (M, N) be a paracompact C^r G-manifold pair, $1 \leq r \leq \omega$, G a discrete group acting on (M, N) such that all the maps $g\colon (M, N) \to (M, N)$ are C^r. Assume that for each $x \in N$ we have $x \in \partial M$ if and only if $x \in \partial N$, and for each $x \in \partial N$, T_xN is transversal to $T_x\partial M$.*

For each component C of M, let $G'_C = H'_C/H_C$, where H'_C is the subgroup of G fixing C, but not necessarily pointwise, and H_C is the subgroup fixing C pointwise.

Then (M, N) has a G-equivariant C^r-triangulation if and only if for all connected components C of M, the action of $G'_C \curvearrowright C$ is properly discontinuous.

Furthermore, if $t_1\colon |K_1, L_1| \to (M, N)$ and $t_2\colon |K_2, L_2| \to (M, N)$ are two different G-equivariant C^r-triangulations, there exists a common G-equivariant subdivision (K_3, L_3) of (K_1, L_1) and (K_2, L_2) and a G-equivariant C^r-triangulation $t\colon |K_3, L_3| \to (M, N)$ such that t differs from t_1 and t_2 by arbitrarily small C^1-changes (see Definition 114 for the exact definition of arbitrarily small*). Furthermore, if $1 \leq r \leq \infty$, then the maps and t, t_1 and t, t_2 are pairwise G-equivariantly C^r-homotopic. If $r = \omega$, the homotopies are of the class C^∞.*

Proof: By Abels-Strantzalos [1], Section 1.4, it follows that H_C is a normal subgroup of H'_C, and the group G'_C is actually well-defined. We remark that H_C does not need to be a normal subgroup of G.

By Theorem 127, the proper discontinuity of each $G'_C \curvearrowright C$ is a necessary condition for the existence of a triangulation.

We prove the sufficiency. It is enough to prove the existence of a triangulation in the special case $M = GC$, where C is connected, since the general case is a disjoint union of such special cases. By Corollary 111, we have a G'_C-equivariant triangulation

$t_C\colon |K_C, L_C| \to (C, N\cap C)$. Since the action of H_C in C is trivial, we can define a simplicial action of H'_C in (K_C, L_C) by $gv = (gH_C)v$ for all $g \in H'_C$, v a vertex of K_C. Then t_C is H'_C-equivariant.

By Corollary 13, we have that $GC = G\times_{H'} C$. Thus, we can define the triangulation of GC by $G\times_{H'} t_C\colon G\times_{H'} (K_C, L_C) \to (G\times_{H'} C, G\times_{H'} (C\cap N)) = (M, N)$. Now we have proved the existence part.

We prove uniqueness. Again, it is enough to consider the case $M = GC$, where C is a connected component, since the general case is a disjoint union of such special cases. Thus, we are given similar coverings $(V_i) \subset\subset (U_i)$ of C with G'_C-coordinate neighbourhoods, and (ϵ_i) , $i \in \mathbb{N}$, $\epsilon_i > 0$ for all i. Note that the neighbourhoods V_i and U_i are H'_C-invariant, and thus they extend into coordinate neighbourhoods GV_i, GU_i of GC such that $GV_i \cap C = V_i$, $GU_i \cap C = U_i$ for all i (even though the action of G might not be properly discontinuous). Thus, we are in the situation of Definition 114.

Now the maps $t_1|t_1^{-1}C$ and $t_2|t_2^{-1}C$ are H'_C-equivariant, and so they can be regarded as G'_C-equivariant, as we define the action of G'_C in the H'_C-complexes $t_1^{-1}C$, $t_2^{-1}C$ by $gH_Cv = gv$. (Since t_1 and t_2 are equivariant homeomorphisms, this action is well-defined.) By Corollary 121, there exist $t'_1|t^{-1}C\colon (K''_1, L''_1) \to (C, N\cap C)$ and $t'_2|t^{-1}C\colon (K''_2, L''_2) \to (C, N\cap C)$ that are (ϵ_i)-approximations to $t_1|$, $t_2|$, which are G'_C-equivariant C^r-triangulations and intersect in a full subcomplex.

Like above, we can regard $t'_1|$, $t'_2|$, (K''_1, L''_1) and (K''_2, L''_2) as H'_C-equivariant maps and H'_C-complexes, respectively. We have that $(K_1, L_1) = (G\times_{H'_C} t_1^{-1}C, G\times_{H'_C} t_1^{-1}C\cap N)$ (and similarly for (K_2, L_2)) by Lemma 139, and thus by Lemma 148, $(K'_1, L'_1) = (G\times_{H'} K''_1, G\times_{H'} L''_1)$ is a G-equivariant subdivision of (K_1, L_1) (and similarly for (K_2, L_2)). Now the maps $t'_1 = G\times_{H'} t'_1|\colon (G\times_{H'} K''_1, G\times_{H'} L''_1) \to (M, N)$ and $t'_2 = G\times_{H'} t'_2|\colon (G\times_{H'} K''_2, G\times_{H'} L''_2) \to (M, N)$ are the required triangulations. They intersect in a subcomplex by Lemma 144.

Since t'_1 and t'_2 are bijective maps that intersect in a subcomplex, we can identify (K'_1, L'_1) and (K'_2, L'_2) via the G-equivariant simplicial isomorphism ${t'_2}^{-1} \circ t'_2$, and obtain the common subdivision $(K_3, L_3) = (K'_1, L'_1) = (K'_2, L'_2)$, and $t' = t'_1 = t'_2$.

We prove that the maps t_1 and t'_1 (resp. t_2 and t'_2) are G-equivariantly homotopic. Again, it is enough to consider the case $M = GC$. By Theorem 120, there is a G'_C-equivariant homotopy $F\colon (|K''_1|, |L''_1|) \times \mathbb{R} \to (C, C\cap N)$ between $t_1|$ and $t'_1|$ of the required differentiability class. By Corollary 13, $(|K'_1| \times \mathbb{R}, |L'_1| \times \mathbb{R}) = G\times_{H'_C} (|K''_1| \times \mathbb{R}, |L''_1| \times \mathbb{R})$, where G acts trivially on $\mathbb{R}$, and thus F extends to a G-equivariant homotopy $G\times F\colon (|K'_1| \times \mathbb{R}, |L'_1| \times \mathbb{R}) \to (M, N)$. Since extensions in the twisted product are unique, $F(\cdot, 0) = t_1$ and $F(\cdot, 1) = t'_1$. A similar argument works also for t_2 and t'_2. □

Acknowledgements

This is my own work.

Appendix A

Twisted product of abstract complexes

Let K be a locally finite simplicial complex, G a discrete group and H a subgroup of G. Let, furthermore, H act on K simplicially. We form the complex $G \times K$, where simplices are of type $g \times s$, where $g \in G$, and s is a simplex of K.

We define a simplicial action of H in $G \times K$ by $h(g,x) = (gh^{-1}, hx)$. We let the quotient $(G \times K)/H$ be the twisted product $G \times_H K$, and let $\Pi \colon G \times K \to G \times_H K$ be the quotient map. The simplicial structure of $G \times_H K$ is determined so that Πs is a simplex of $G \times_H K$ for each simplex s of $G \times K$.

Lemma 129 *We have the following:*

1. *We have the natural H-equivariant embedding $i \colon K \to G \times_H K$.*

2. *If s is an n-simplex of $G \times K$, then Πs is an n-simplex of $G \times_H K$.*

Proof: The map i is injective and H-equivariant, since if we look at only vertices of the complexes and forget the simplicial structure, our definition is essentially the topological twisted product of discrete spaces. That i is simplicial follows directly from the definition. This proves (1).

Let s be a n-simplex of $G \times K$. By definition, Πs is a simplex, and all we have to check is that Π maps the vertices of s injectively. But each vertex of s is of the form (g,x) for a fixed g, and such points are not identified by the quotient operation. This proves (2). □

We define an action of G in $G \times_H K$ by $g[g', x] = [gg', x]$.

Lemma 130 *The action is well-defined and simplicial.*

Proof: That it is well-defined follows by the theory of the topological twisted product, as in the previous lemma. We show that the action is simplicial. Let $g \in G$, and let

$s = \Pi((g', x_1), \ldots, (g', x_n))$ be a simplex of $G \times_H K$. Then $gs = \Pi((gg', x_1), \ldots, (gg', x_n))$ is also a simplex of $G \times_H K$.□

Let K be a G-complex and L an H-complex. Let f be an H-equivariant simplicial map $L \to K$. Then there exists a map $G \times_H f \colon G \times_H L \to K$ defined by $f([g, v]) = gf(v)$.

Lemma 131 *$G \times_H f$ is well-defined, G-equivariant and simplicial.*

Proof: That $G \times_H f$ is well-defined and G-equivariant follows by the theory of the topological twisted product when applied to the discrete spaces of vertices. We show that G is simplicial. Let $g \in G$, and let $s = ([g, v_1], \ldots, [g, v_n])$ be a simplex in $G \times_H L$. Now $s' = (v_1, \ldots, v_n)$ is a simplex in L. Thus fs' is a simplex of K, and so $gfs' = (G \times_H f)s$ is a simplex of K. □

Now we know the basic properties of the twisted product of abstract complexes.

We define next an analogue of non-connectedness for complexes and study it a little.

Definition 132 *Let L be a subcomplex of a simplicial complex K. We say that L is isolated, if the following holds: If s is a simplex of K such that $s \cap L \neq \emptyset$, then s is a simplex of L.*

Lemma 133 *L is isolated in K if and only if $|L|$ is closed and open in $|K|$.*

Proof: *if* The subset $|L|$ is always closed in $|K|$. Let $x \in |L|$. Then $\bigcup_{x \in |s|} |s|$, is a neighbourhood of x in $|L|$. Write $x = \sum t_i v_i$, $t_i > 0$. Since $x \in |L|$, each $v_i \in L$. But $x \in |s|$ is equivalent to the claim that all v_i's are vertices of s, and by the isolatedness condition s is a simplex of L. Thus, $\bigcup_{x \in |s|} |s|$ is a neighbourhood of x, which is contained in $|L|$. Thus, $|L|$ is open in $|K|$. We remark that this argument works both in the metric topology and the weak topology of $|K|$.

only if Assume L is not isolated in K. Let v be a vertex of L and $s \ni v$ a simplex of K, but not a simplex of L. It follows that no inner point of $|s|$ is a point of $|L|$. All neighbourhoods of $v \in |K|$ contain inner points of $|s|$, and thus $|L|$ is not open in $|K|$. We remark that also this argument works both in the metric topology and the weak topology.□

Lemma 134 *L is isolated in $G \times_H L$.*

Proof: Let s be a simplex of $G \times_H L$. Thus, $s = ([g, v_1], \ldots, [g, v_n])$, where $g \in G$ and v_i's are vertices of L. If $g \in H$, then s is a simplex of L. If $g \notin H$, then none of the vertices gv_i are vertices of L. Thus no simplex of $G \times_H L$ has vertices both in L and in $K \setminus L$. □

Next, we study the connection between the simplicial twisted product and the topological twisted product.

Lemma 135 *The natural H-equivariant imbedding $i\colon K \hookrightarrow G \times_H K$ induces an H-equivariant map $|i|\colon |K| \hookrightarrow |G \times_H K|$, and a G-equivariant map $G \times_H |i|\colon G \times_H |K| \to |G \times_H K|$.*

Furthermore $G \times_H |i|$ is an isomorphism.

Proof: The first part is obvious, and we only need to prove that $G \times_H |i|$ is an isomorphism.

Thus, we need to prove that $|i||K|$ satisfies the assumptions of Lemma 11. We have that iK is isolated in $G \times_H K$ by Lemma 134. Hence, $|i||K|$ is open in $|G \times_H K|$.

Let s be a simplex of K. Since $gs \subset K \subset G \times_H K$ if and only if $g \in H$, we have that $|gs| \subset |K| \subset |G \times_H K|$ if and only $g \in H$. Since $|gs| \subset |K|$ or $|gs| \cap |K| = \emptyset$, if follows that $\sum t_i g v_i \in |K|$ (v_i's vertices of a simplex of K) if and only if $g \in H$. Furthermore, $G|K| = |G \times_H K|$, since if we are given a point $\sum t_i(gv_i)$ of $|G \times_H K|$ (note that all simplices of $G \times_H K$ are of the form $g(v_1 \dots v_n)$, where v_i's are points of K, and thus each point of $|G \times_H K|$ can be represented as $\sum t_i(gv_i)$), then $G \times_H |i|(g \sum t_i v_i) = \sum t_i(gv_i)$. Thus, the assumptions of Lemma 11 are satisfied, and $G \times_H |i|$ is a G-equivariant isomorphism.□

Remark 136 *We will often identify $G \times_H |K|$ and $|G \times_H K|$ through the isomorphism of the previous lemma. Note that $|K| \hookrightarrow |G \times_H K|$ is simplexwise linear and the image of a simplex is a simplex, and thus $G \times_H |K|$ and $|G \times_H K|$ can be thought to have the same simplexwise linear structure.*

Corollary 137 *If $t\colon |K| \to X'$ is an $G_{X'} = H$-equivariant map, where X' is a special component of a generalized coordinate neighbourhood X, t extends uniquely to a G-equivariant map $G \times_H t\colon |G \times_H K| \to X$.*

If the original map t is differentiable (immersion), also $G \times_H t$ is differentiable (immersion).

Proof: To prove the first part we note that we have the extension $|G \times_H K| \overset{\cong}{\to} G \times_H |K| \overset{G \times_H t}{\to} X$, and in the continuation, we identify also this composed map with $G \times_H t$.

We prove the last part. Let $|s| = |([g, x_1], \dots, [g, x_n])|$ be a simplex of $|G \times_H K|$. Let $|s'| = g^{-1}|s| \subset |K|$. Now $t' = G \times_H t|\,|s| = gtg^{-1}|\,|s|$, where g^{-1} is a simplexwise linear map. Thus t' is differentiable (immersion) if and only if $gt|\,|s'|$ is differentiable (immersion), but this map is differentiable (immersion) as a composition of two differentiable (immersion) maps.

To complete the proof about $G \times_H t$ being an immersion, we must prove that the angle between the images of two simplices gs, $g's'$ having a common face $g''s''$, is positive. From the definition of $G \times_H K$, we deduce easily that we can choose in the above presentation $g = g' = g''$. But now $G \times_H t|\,|gs \cup gs'| = g \circ (t|\,|s \cup s'|) \circ g^{-1}$, where g^{-1} is a simplexwise linear isometry, and the two other component functions are immersions. Thus, the angle between s and s' is positive. □

An invariant subdivision K' of K is a G-complex together with a G-equivariant homeomorphism $t\colon |K'| \to |K|$ such that for each simplex s of K', we have that t maps $|s|$ linearly inside $|s'|$, where s' is a simplex of K.

Lemma 138 *If K' is an H-invariant subdivision of K, then $G \times_H K'$ is a G-equivariant subdivision of $G \times_H K$.*

Proof: We have the natural simplexwise linear homeomorphism $h\colon |G \times K'| \to |G \times K|$, and this gives us the required simplexwise linear equivariant homeomorphism if we take the quotients by H, since $G \times K'$ and $G \times K$ are regular H-complexes in the sense of Kawakubo [13], Chapter 5. □

We prove the following analogue of Lemma 11.

Lemma 139 *Let K be a G-complex, $H < G$, and L be a subcomplex of K satisfying the following*

1. *L is isolated in K*
2. *$GL = K$.*
3. *Let v be a vertex of K. Then $hv \in L$ if and only if $v \in L$ for all $h \in H$.*

Then the natural embedding $i\colon L \to K$ induces a G-equivariant isomorphism $G \times_H i \to G \times_H L \to K$.

Proof: We apply Lemma 11 to the discrete spaces of vertices of L, $G \times_H L$ and K. This gives us the result that $G \times_H i$ is a bijection.

We need to prove that $G \times_H i$ is a simplicial isomorphism. By Lemma 131, it is a simplicial map.

Let s be a simplex of K and v a vertex of s. Since $GL = K$, there exists $g \in G$ such that $gv \in L$. Since L is isolated, we must have that gs is a simplex of L. But now $g^{-1} G \times_H i(gs) = G \times_H i(g^{-1}gs) = s$. Thus each simplex s of K is an image of a simplex of $G \times_H L$. □

Corollary 140 *Let K be a G-complex and L be an H-invariant subcomplex of K. Then $K = G \times_H L$ if and only if L satisfies the numbered conditions of the previous lemma.*

Proof: The previous lemma is the *if*-direction. We prove the *only if*-direction. Condition (1) is the Lemma 134. Condition (3) follows from Corollary 12, applied to the discrete spaces of vertices. (2) follows from the definition of the twisted product. □

Corollary 141 *Assume $K = G \times_H L$, and L' is an H-invariant subcomplex of L. Then $G \times_H L' = GL' \subset K$.*

Proof: By the previous corollary, the pair L, K satisfies the assumptions of Lemma 139. The assumptions are easily seen to be hereditary in the sense that the pair L', GL' satisfies them, too. Thus, $GL' = G \times_H L'$. □

Corollary 142 *Let K be a G-complex and L its H-invariant subcomplex. Denote the natural inclusion by $i\colon L \to K$. Then $G \times_H i\colon G \times_H L \to K$ is an isomorphism if and only if $G \times_H |i|\colon G \times_H |L| = |K|$ is an isomorphism.*

Proof: If $G \times_H i\colon G \times_H L \to K$ is an isomorphism, then $|i|, |K|$ satisfy the assumptions of Lemma 11, and thus $G \times_H |L| = |K|$.

If $|i|\colon G \times_H |L| \to |K|$ is an isomorphism, then i, K satisfy the assumptions of Lemma 139, and thus $G \times_H L = K$.□

Corollary 143 *If $f\colon |K| \to X$ is G-equivariant, and X is a coordinate space, then $K = G \times_{G_{X'}} L$, where L is the subcomplex of K whose geometrical realization is $f^{-1}X'$.*

Proof: By Corollary 14 we have that $|K| = G \times_{G_{X'}} |L|$, and by the previous corollary we have that $K = G \times_{G_{X'}} L$.□

We next prove that taking the twisted product preserves the intersecting in a subcomplex property.

Lemma 144 *Let K and L be $G_{X'}$-complexes and $f\colon |K| \to X'$ and $f'\colon |L| \to X'$ be $G_{X'}$-invariant embeddings to the special component of a coordinate space X.*

If f and f' intersect in a subcomplex (resp., full subcomplex) , then $G \times_{G_{X'}} f$ and $G \times_{G_{X'}} f'$ intersect in a subcomplex (resp., full subcomplex).

Proof: Let $H = G_{X'}$. Let $F = G \times_H f$ and $F' = G \times_H f'$.

Assume that $F|g\dot{s}|$ and $F'|g's'|$ intersect, where s is a simplex of K and s' is a simplex of L. We make the antithesis that $F'^{-1}F|\,|gs|$ is not an affine isomorphism of $|gs|$ and $|g's'|$.

For a point $g'x \in |g's'| \cap |gs|$, we have that $g^{-1}F'(g'x) \in X'$, and thus $h = g^{-1}g' \in H$. Now $f'|h\dot{s}'| = g^{-1}F'|g'\dot{s}'|$ and $f|\dot{s}| = g^{-1}F|g\dot{s}|$ intersect, but $f'^{-1}f|\,|s|$ is not an affine isomorphism of $|s|$ and $|hs'|$. This is a contradiction with the assumption that f and f' intersect in a subcomplex.

The claim about full subcomplexes follows, as one proves that if K_0 is an H-invariant full subcomplex of K, then GK_0 is a full subcomplex of $G \times_H K$. Let s be a simplex of GK such that all vertices v_i of s are in GK_0. Choose g such that $gs \cap K \neq \emptyset$. Since K

is isolated, we must have $gs \subset K$. Since GK_0 is G-invariant, also each vertex of gs is in GK_0. One easily sees that $GK_0 \cap K = K_0$, and so each vertex of K_0 is in K_0. Since K_0 is a full subcomplex of K and gs is a simplex of K, we have that gs is a simplex of GK_0. Thus $s = g^{-1}gs$ is a simplex of GK_0. The proof is complete. □

The rest of this section deals with various ways to glue together subdivisions of a part of the complex to obtain a subdivision of the entire complex.

Definition 145 *Let K be a simplicial complex and L a subcomplex of K. We define that ∂L is the G-invariant subcomplex of L consisting of all simplices s of L such that s is a face of a simplex not in L.*

We define that $K - L$ is the subcomplex of K consisting of all simplices s of K such that s is a face of a simplex not in L.

Remark 146 *The following are immediate from the definition:*

- *If K is a G-complex and L is G-invariant, then ∂L and $K - L$ are G-invariant.*
- *$L \cap (K - L) = \partial L$.*
- *Each simplex of K is a simplex of L or $K - L$.*

Proposition 147 *Let K be a G-complex and L a G-invariant subcomplex of K. Assume L' is a G-invariant subdivision of L such that the simplicial structures of L and L' coincide in the boundary. To be exact, the last condition means that there exists a diagram*

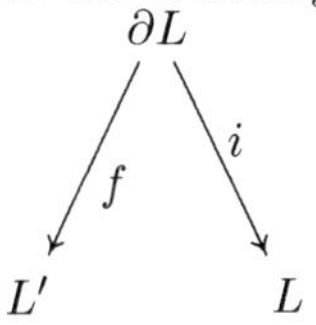

where f is a G-equivariant embedding and i is the canonical embedding such that the diagram gives rise to a commutative diagram

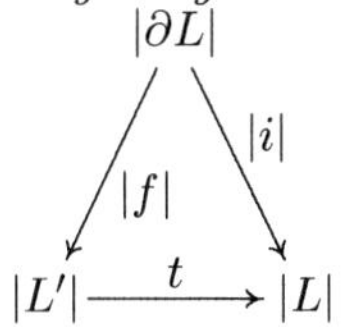

where t is the subdivision map.

Claim: *Then we get an equivariant subdivision K' of K such that the simplicial structure of K' coincides with that of $K - L$ outside L and with that of L' inside L. To be exact, these claims mean that the diagrams*

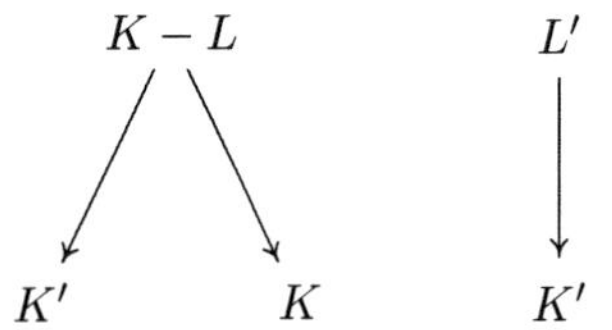

give rise to commutative diagrams

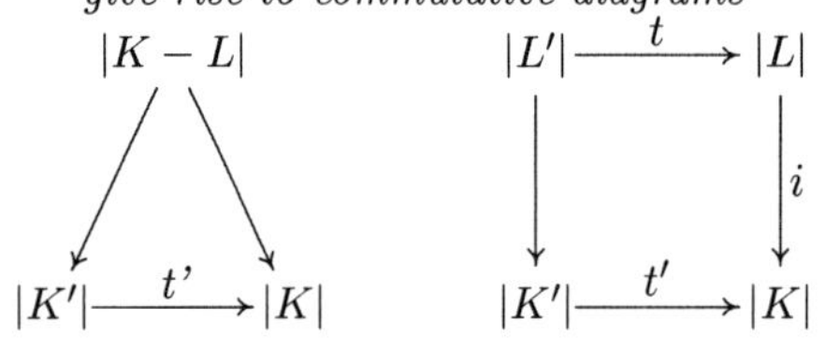

such that all unnamed maps are G-equivariant embeddings, and t' is an equivariant subdivision map.

Proof: First, we form the simplicial complex $K'' = L' \coprod (K - L)$. Then, we form the complex $K' = L' \coprod_{\partial L} (K - L)$ by identifying vertices $v' \in L'$ and $v \in K - L$ if and only if $f(v) = v'$. The simplicial structure of K' is determined so that simplices of K' are the images of simplices of K'' under the canonical projection $K'' \to K'$.

We claim that $g[v] = [gv]$ defines an action of G in K'. If $v \neq v'$, $[v] = [v']$, then $v' \in L'$, $v \in K - L$ (or vice versa), and $fv = v'$. Since f is G-equivariant, we have that $fgv = gv'$, and thus $[gv] = [gv']$. Thus the action of G is well-defined. The action is simplicial, since if s is a simplex of K', then $s = [s']$, where s' is a simplex of either L' of $K - L$. Thus gs' is a simplex of L' or $K - L$, and $gs = [gs']$ is a simplex of K'.

Next, we construct the map $t'|\colon K' \to |K|$ in the set of vertices of K' by $t'(v) = t(v)$, if $v \in L'$, and $t'(v) = v$, if $v \in K - L$. The map t is well defined, since if $[v'] = [v]$, $v' \in L'$, $v \in K - L$, then $v' = fv$, and by the commutativity of the first diagram in the statement of the proposition $tv' = v$.

The map t' is G-equivariant, since t and the canonical imbedding $K - L \to K$ are G-equivariant.

Next, we want to extend t' into a map $t'\colon |K'| \to |K|$ linearly. If we succeed in this, t' is obviously G-equivariant, and makes the two diagrams in the statement of the proposition commutative. We succeed in the extension if and only if for each simplex s of K' we find a simplex s' of K such that t' maps all the vertices of s into $|s'|$.

Let s be a simplex of K'. Then s is either a simplex of $K - L$ or L'. In the former case, $t's = s$, and in the above condition we can choose $s' = s$. In the latter case, there is a simplex s' of L such that t maps vertices of s into $|s'|$, since t is a subdivision map. Thus t' maps the vertices of s into $|s'|$, and we are done.

Thus there exists $t'\colon |K'| \to |K|$ making the required diagrams commutative. We must still check that t' is a subdivision map, that is, bijective. We check surjectivity first.

Let $\sum t_i v_i$ be a point in $|K|$, v_i's vertices of a simplex s of K. If s is in $K-L$, then $t'(\sum t_i v_i) = \sum t_i v_i$. If not, then s is a simplex of L and there exists $\sum t'_i v'_i \in |L'|$ such that $t(\sum t'_i v'_i) = \sum t_i v_i$, and thus $t'(\sum t'_i v'_i) = t_i v_i$. Thus t' is surjective.

We check that t' is injective. We immediately see that $t'||K-L|$ and $t'|L'|$ are injective. Thus we are left with the situation where $\sum t_i v_i \in |L'|$, $\sum t'_i v'_i \in |K-L|$, $t_i, t'_i > 0$, $t(\sum t_i v_i) = t'(\sum t_i v_i) = t'(\sum t'_i v'_i) = \sum t'_i v'_i$. Since $t(\sum t_i v_i) \in |L|$, $\sum t'_i v'_i \in |K-L|$, we have that $t'_i v'_i \in |L| \cap |K-L| = |\partial L|$. Thus, $|t||f| \sum t'_i v'_i = |i| \sum t'_i v'_i = |t'| \sum t'_i v'_i = \sum t'_i v'_i$, and since $|t|$ is injective, $|f| \sum t'_i v'_i = \sum t_i v_i$. But $|f| \sum t'_i v'_i = \sum t'_i f v'_i$, and since this representation is unique up to indexing i, we have after re-indexing that $t_i = t'_i$, $f v'_i = v_i$ for all i. But this means that $[f v'_i] = [v_i]$, and, thus, $\sum t_i v_i = \sum t'_i v'_i$, and t' is injective.□

Proposition 148 *Let K be a G-complex, and L an H-invariant subcomplex such that $G \times_H L = GL$. Assume that L' is an H-invariant subdivision of L such that the simplicial structures of L' and K coincide on ∂L. Then there exists a G-invariant subdivision K' of K such that the simplicial structure of K' coincides with the simplicial structure of L' inside L and with the simplicial structure of K outside GL.*

Proof: We have to check that K and $G \times_H L'$ satisfy the assumptions of the previous proposition. Then the claim follows from the previous proposition.

Denote by t the subdivision map $|L'| \to |L|$.

First, we note that since the subdivision map $t\colon |L'| \to |L|$ is bijective and simplexwise linear, also the map $G \times_H t\colon |G \times_H L'| = G \times_H |L'| \to G \times_H |L| = |GL|$ is bijective and simplexwise linear, that is, a subdivision map.

Next, we prove that $\partial GL = G\partial L$. If s is a simplex of ∂GL, then s is a simplex of GL, and s is a face of a simplex s' not in GL. Thus we can choose $g \in G$ such that gs is a simplex of L, and gs' is a simplex with gs as a face such that gs' is not a simplex of GL. So gs' is not a simplex of L, and, consequently, gs is a simplex of ∂L, and s is a simplex of $G\partial L$. Hence ∂GL is a subcomplex of $G\partial L$.

Assume s is a simplex of $G\partial L$. Choose $g \in G$ such that $gs \cap \partial L \neq \emptyset$. Since L is isolated in $GL = G \times_H L$, ∂L is isolated in $G\partial L$, and, thus, gs, as a simplex of $G\partial L$, is a simplex of ∂L. Let gs' be a simplex of K but not of L such that gs is a face of gs'. We claim that gs' is not a simplex of GL. If it is a simplex of GL, then gs' is a simplex of GL such that gs' has vertices both in L (namely, those of gs), and in $GL \setminus L$. However, this is impossible, since L is isolated in $G \times_H L = GL$. Thus gs' is not a simplex of GL. But now s' is a simplex of K, but not of GL, with a face s in GL. Thus, s is a simplex of ∂GL. Hence, $G\partial L$ is a subcomplex of ∂GL.

Putting together the results in the previous two paragraphs, we see that $\partial GL = G\partial L$. Since ∂L is a subcomplex of L, we have by Corollary 141 that $\partial GL = G\partial L = G \times_H \partial L$.

By the assumption of the theorem, there is an H-invariant map $f\colon \partial L \to L'$ making the diagram

$$\begin{array}{ccc} & |\partial L| & \\ {}^{|f|}\swarrow & & \searrow^{|i|} \\ |L'| & \xrightarrow{\quad t \quad} & |L| \end{array}$$

commutative. Thus, applying the $G\times_H$ functor to the diagram we get the commutative diagram

$$\begin{array}{ccc} & |G\times_H \partial L| = |\partial GL| & \\ {}^{|G\times_H f|}\swarrow & & \searrow^{|G\times_H i|} \\ |G\times_H L'| & \xrightarrow{\quad G\times_H t \quad} & |GL| \end{array}$$

satisfying the assumptions of the previous proposition. Note that $G\times_H i$ equals the canonical inclusion of ∂GL into GL. □

Proposition 149 *Assume s is a G-equivariant simplex and $\overset{\bullet}{s}'$ is a G-equivariant subdivision of $\overset{\bullet}{s}$. Let z be a simplex of s, and let $H < G$ be the group G_z. Let, furthermore, z' be an H-equivariant subdivision of z such that the simplicial structures of z' and $\overset{\bullet}{s}'$ coincide in $\overset{\bullet}{z}$. Then z' and $\overset{\bullet}{s}'$ extend to a G-equivariant subdivision s' of s.*

Proof: To avoid conflicting notation, denote $q = \overset{\bullet}{s}'$ in the formulation of the proposition. To simplify the notation, identify $|q|$ with a subset of $|\overset{\bullet}{s}|$ of $|s|$.

Let $s'' = G\times_H z'$, and let $t\colon |z'| \to |z|$ be the subdivision map. Form $f = G\times_H t\colon |s''| \to |s|$, and let s' be the complex obtained from s'' by identifying vertices $v \sim v'$ if and only if $fv = fv'$. Simplices of s' are the projections of simplices of s''.

We claim that s' inherits an action of G from s''. We prove that it is well-defined. Thus, assume that $v \sim v' \in s''$. Then $f(v) = f(v')$, and consequently, $f(gv) = gf(v) = gf(v') = f(gv')$, that is, $gv \sim gv'$. Thus the action is well-defined. We easily see that the action is simplicial. Denote $\overset{\bullet}{s}'' = (G\times_H t)^{-1}\overset{\bullet}{s}$, and $\overset{\bullet}{s}' = \operatorname{pr} \overset{\bullet}{s}''$.

Let $t'|\colon s' \to |s|$, defined on the vertices of s', be the map such that $t'[v] = f(v)$ for all vertices v of s'. We readily see that $t'|$ is G-equivariant. Now, let $[r]$ be an arbitrary simplex of s', where r is a simplex of s''. Then, $r = gr'$ for r' a simplex of z', and thus $f|r = gtg^{-1}|r$, and we easily see that $\operatorname{im} gtg^{-1}|r$ is contained in a simplex of s. Thus, $\operatorname{im} t'|[r] = \operatorname{im} f|r$ is contained in a simplex of s. Thus, $t'|$ extends linearly into a G-equivariant map $t'\colon |s'| \to |s|$.

We claim that t' is the required subdivision map. To complete the proof, we must show the following

1. t' is surjective.

2. t' is injective.

3. t' and t intersect in a subcomplex.

4. t' and $t''\colon |q| \to |s|$ intersect in a subcomplex.

Proof of 1: It is enough to see that f is surjective, but this is true, since the domain of f is $G \times_H |z|$, and the range $s = G|z|$. Thus (1) is complete.

We have the diagram

$$s' \longleftarrow G \times_H z' \longleftarrow z'$$

giving rise to a commutative diagram

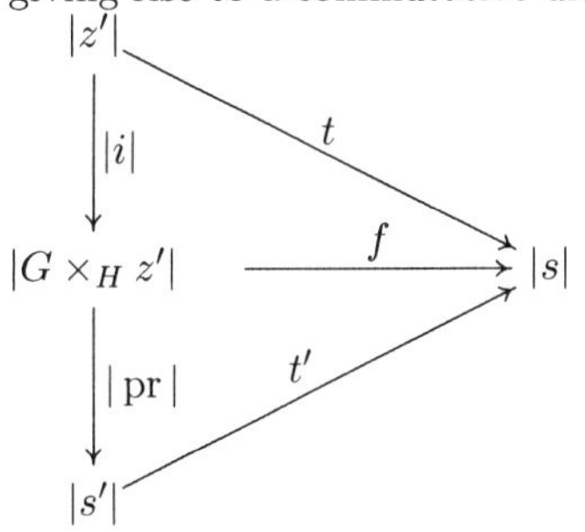

Since t is injective, we have that $t'|g \operatorname{im} |\operatorname{pr} \circ i|$ is injective for all $g \in G$. Furthermore, if t' is (globally) injective, t and t' intersect in a subcomplex; in that case, $t'^{-1} \circ t$ is, namely, the simplicial map $|i \circ \operatorname{pr}|$. So we are left with (2) and (4).

We prove (2). Let $z'' = \operatorname{im}(\operatorname{pr} \circ i)$. We easily see that $|\overset{\circ}{z}|$ and $|\overset{\circ}{s}|$ satisfy the assumptions of Lemma 11, and thus $|\overset{\circ}{s}| = G \times_H |\overset{\circ}{z}|$, and $f|\, G \times_H |\overset{\circ}{z}|$ is an isomorphism, in particular, injective. Thus $t'|G|\overset{\circ}{z''}|$ is injective.

Next we claim that $t'|\colon \overset{\bullet}{s'} \to q$ is a simplicial injective map. We have that $t'v = f \operatorname{pr}^{-1}(v)$ for all vertices v of $\overset{\bullet}{s'}$, and $\operatorname{pr}^{-1}(v)$ is a vertex of $G \times_H \overset{\bullet}{z'}$ (for any choice of $\operatorname{pr}^{-1} v$). Furthermore, f maps a vertex of $\overset{\bullet}{z'}$ into a vertex of q (by assumption), and since q is a G-complex, f maps a vertex of $G \times_H \overset{\bullet}{z'}$ into a vertex of q. So t' maps a vertex of $\overset{\bullet}{s'}$ into a vertex of q. If $t'|[v] = t'|[v']$, then $f(v) = f(v')$, and thus $[v] = [v']$. Thus, $t'|$ is injective in the set of vertices of $\overset{\bullet}{s'}$. Let $[gz''']$ be an arbitrary simplex of $\overset{\bullet}{s'}$, where gz''' is a simplex of $\overset{\bullet}{s''}$, and z''' is a simplex of $\overset{\bullet}{z'} \subset q$. Thus, gz''' is a simplex of q, and $t'|$ maps the abstract simplex $[qz''']$ onto an abstract simplex of q. Therefore $t'|$ is a simplicial injective map of abstract complexes $\overset{\bullet}{s'} \to q$. So, $t'|\,|\overset{\bullet}{s'}| \to |q|$ is injective, and $t'|\,|\overset{\bullet}{s'}|$ and $|q|$ intersect in a subcomplex.

Now t' is injective, since $\operatorname{im} t'|\,|\overset{\bullet}{s'}| = |\overset{\bullet}{s}|$, and $\operatorname{im} t'|\,|\overset{\circ}{s'}| = |\overset{\circ}{s}|$ do not intersect, and both restrictions of t' are injective. Thus we have (2).

t' and q intersect in a subcomplex, since $t'|\,|\overset{\bullet}{s'}|$ and q intersect in a subcomplex, and $t'|\,|\overset{\circ}{s'}|$ and q do not intersect. Thus we have (4). □

Corollary 150 *Assume K is an n-dimensional simplicial G-complex. Assume K'^{n-1} is a G-equivariant subdivision of K^{n-1}. For each equivariant n-simplex s_j of K, let z_j be an n-simplex in s. Let z_j' be a G_z-equivariant subdivision of z_j such that the simplicial structures of z_j' and z_j coincide on $\overset{\bullet}{z}$. Then z_j' and K'^{n-1} extend to a G-equivariant subdivision of K'.*

Proof: Let $K_j = K^{n-1} \cup s_j$. Then the previous proposition and Proposition 147 give us a subdivision K_j' of K_j such that K_j' coincide with K'^{n-1} and z_j'. Furthermore, the natural embeddings $i_j\colon |K_j| \to |K|$ intersect in the subcomplex K^{n-1}, and the natural embeddings $i_j'\colon |K_j'| \to |K|$ intersect in the subcomplex K'^{n-1}.

In the terminology of Section 4.1, we have that $K = \coprod/K_j$; the map $\coprod/i_j$ is well-defined, since it equals the natural identification map $\coprod/K_j = K$. By Proposition 105, we have that $K' = \coprod/K_j'$ is an equivariant subdivision of K.□

Acknowledgements

This is my own work, although I had the theory of the topological twisted product as a model. The main use for these theorems is to have a mathematically exact way to deal with different subdivisions that arise when approximating maps from simplicial complexes into manifolds. However, here my contribution is not in the techniques but in proving that they work. Many of the techniques, such as Proposition 149, were used implicitly by Illman [7].

Bibliography

[1] Abels, Herbert and Strantzalos, Polychronis, *Proper Transformation Groups*, manuscript, August 2001

[2] Bochner, Salomon, *Compact Groups of Differentiable Transformations*, Ann. of Math. 46(1945), 372-381

[3] Bredon, Glen E., *Introduction to Compact Transformation Groups*, Academic Press, Inc., Orlando/ San Diego/ New York/ London/ Toronto/ Montreal/ Sydney/ Tokyo 1972

[4] Hironaka, Heisuke, *Triangulations of Algebraic Sets*, Proc. Sympos. Pure Math., Algebraic Geometry, Arcata 1974, Am. Math. Soc. 29 (1973), 165-185

[5] Giesecke, B., *Simpliziale Zerlegung abzõhlbarer analytischer Rõume*, Math. Z. 83 (1964), 177-213

[6] Hirsch, Morris W. and Mazur, Barry, *Smoothings of Piecewise Linear Manifolds*, Annals of Mathematics Studies 80, Princeton University Press, 1974

[7] Illman, Sören, *Smooth Equivariant Triangulations of G-Manifolds for G a Finite Group*, Math. Ann. 244(1978), 199-220

[8] Illman, Sören, *Every proper smooth action of a Lie group is equivalent to a real analytic action: a contribution to Hilbert's fifth problem*, Ann. of Math. Stud. 138 (1995), 189-220

[9] Illman, Sören, *Existence and Uniqueness of Equivariant Triangulations of Smooth Proper G-manifolds with Some Applications to Equivariant Whitehead Torsion*, J. Reine Angew. Math. 524 (2000), 129-183

[10] Illman, Sören and Kankaanrinta, Marja, *A New Topology for the Set* $C^{\infty,G}(M,N)$ *of G-Equivariant Smooth Maps*, Math. Ann. 316(1), 2000, 139-168.

[11] Kankaanrinta, Marja, *Actions of Lie Groups on Manifolds*, Ann. Acad. Sci. Fenn. Ser. A. I Math. Dissertat. 83 (1991)

[12] Kaup, L and Kaup, B: *Holomorphic Functions of Several Variables*, Walter de Grynter, Berlin / New York, 1983.

[13] Kawakubo, Katsuo, *The Theory of Transformation Groups*, Oxford University Press, Oxford/ New York/ Tokyo, 1991

[14] Kelley, J.L., *General Topology*, D Van Nostrand Company, Inc., Princeton/ Toronto/ London/ New York, 1955

[15] Korppi, Tuomas, *Equivariant Triangulations of C^r-Manifolds with a Properly Discontinuous Action*, Licentiate thesis, University of Helsinki, 2002.

[16] Lojasiewics, S. *Triangulation of Semi-Analytic sets*, Ann. Scu. Norm. Sup. Pisa (3) 18 (1965), 449-474

[17] Matumoto, Takao and Shiota, Masahiro, *Proper Subanalytic Transformation groups and Unique Triangulation of the Orbit Space*, Transformation Groups, Poznan 1985, Proceedings, Springer Lect. Notes Math. 1217 (1986), 290-302

[18] Munkres, James R., *Elementary Differential Topology*, Princeton University Press, Princeton/ New Jersey, 1963

[19] Narasimhan, Raghavan, *Analysis on Real and Complex Manifolds*, North Holland Publishing Company, Paris/ Amsterdam, 1968

[20] Pears, A. R., *Dimension Theory of General Spaces*, Cambridge University Press, Cambridge/ London/ New York/ Melbourne, 1975

[21] Rourke, C.P., Sanderson, B.J., *Introduction to Piecewise Linear Topology*, Springer–Verlag, Berlin/ Heidelberg / New York, 1972

[22] Shiota, Masahiro, *Piecewise Linearization of Real Analytic Functions*, Publ. R.I.M.S. Kyoto Univ. 20, 4 (1984), 727-792

[23] Spanier, Edwin H. *Algebraic Topology*, 2nd Springer Printing, New York/ Berlin/ Heidelberg/ Barcelona/ Budapest/ Hong Kong/ London/ Milan/ Paris/ Singapore/ Tokyo (Originally published by McGraw-Hill 1966)

[24] Whitehead, J.H.C., *On C^1-complexes*, Ann. of Math. 41 (1940), 809-824.